Wheat Blast

Wheat Blast

Edited by

Sudheer Kumar, Prem Lal Kashyap, and
Gyanendra Pratap Singh

CRC Press
Taylor & Francis Group
Boca Raton London New York

CRC Press is an imprint of the
Taylor & Francis Group, an **informa** business

First edition published 2020
by CRC Press
6000 Broken Sound Parkway NW, Suite 300, Boca Raton, FL 33487-2742

and by CRC Press
2 Park Square, Milton Park, Abingdon, Oxon, OX14 4RN

Library of Congress Cataloging-in-Publication Data

Names: Kumar, Sudheer, 1972- editor. | Kashyap, Prem Lal, 1981- editor. | Gyanendra Singh, editor.
Title: Wheat blast / edited by Sudheer Kumar, Prem Lal Kashyap, and Gyanendra Pratap Singh.
Description: Boca Raton, FL : CRC Press, [2020] | Includes bibliographical references and index.
Identifiers: LCCN 2020009491 | ISBN 9781138599970 (hardback) | ISBN 9780429470554 (ebook)
Subjects: LCSH: Wheat--Diseases and pests--Control.
Classification: LCC SB608.W5 W435 2020 | DDC 633.1/1--dc23
LC record available at https://lccn.loc.gov/2020009491

ISBN: 978-1-138-59997-0 (Hardback)
ISBN: 978-0-429-47055-4 (eBook)

Typeset in Times
by Deanta Global Publishing Services, Chennai, India

Contents

Preface

Wheat blast caused by *Magnaporthe oryzae* pathotype *Triticum* (MoT) is one of the most formidable wheat diseases, first reported from Brazil in 1985, widespread in South American wheat fields and now expanded to Asia. It has the potential to cause substantial yield losses of up to 100% under favourable environmental conditions. Its recent outbreak in South Asia, especially in Bangladesh, poses a threat to wheat production in Asia and challenges whcat researchers to manage the disease by identifying sources of resistance and developing resistant varieties, to elucidate the epidemiology of the disease and to find optimal control practices for wheat blast.

Unfortunately, the fungus is physiologically and genetically complex, so even after more than three decades, researchers do not fully understand how it interacts with wheat or which genes can confer durable resistance. The research efforts to understand and manage the disease remain restricted to the countries in which it prevails. There are a number of reviews and publications on specific aspects of wheat blast, but nowhere is all detailed research on the wheat blast menace summarised in one publication. This book provides a review of the research progress made in wheat blast. It provides practical, up-to-date information on wheat blast pathology, summarises research progress and discusses future perspectives based on current understanding of the existing issues and advancing technologies that may help in deciding the future path for further research and development of better strategies and techniques to manage the disease.

This book encompasses the latest literature on the precise identification of diseases as well as the pathogen that causes wheat blast, the conditions that lead to the development of epidemics of wheat blast, the pathogenomics of wheat blast and innovative methods of disease management in an eco-friendly manner. It also covers the biology, genetics and functional genomics of the pathogen, diagnostics, resistance mechanisms and the epidemiology of the disease. This book will be highly valuable to researchers, students, teachers, farmers, seed growers, traders and other stakeholders dealing with wheat. Moreover, in the present scenario, wheat disease is a comparatively new disease to South Asia and is continuously expanding in new directions. Therefore, clear and succinct chapters contributed by researchers in their expertise areas of wheat blast will be highly valuable to policy makers, quarantine specialists and other government agencies dealing in the export and import of wheat, to develop strategies to prevent further spread. It will also be a valuable reference book for consultants and administrators in agricultural businesses and education.

Sudheer Kumar
Prem Lal Kashyap
Gyanendra Pratap Singh

Editors

Sudheer Kumar, PhD, is currently working at ICAR – Indian Institute of Wheat and Barley Research, Karnal, Haryana, India. His areas of specialisation are wheat pathology, biocontrol, diagnostics and host–parasite interactions. He earned his master's degree from S.K. Rajasthan Agriculture University, Bikaner, and secured a gold medal. He earned his PhD from C.C.S. Haryana Agriculture University, Hisar, Haryana, and worked on multiple disease resistance in *Brassica* spp. Dr Kumar started his career as Assistant Professor at Sher-e-Kashmir University of Agriculture Sciences and Technology, Jammu, teaching both graduate and postgraduate classes and supervising many MSc and PhD students. Dr Kumar subsequently moved to ICAR and joined the National Bureau of Agriculturally Important Microorganisms, Mau, UP. There, he established the National Agriculturally Important Microbial Culture Collection (NAIMCC) which is recognised as one of the national repositories by the Biodiversity Authority of India. Dr Kumar also worked as a visiting research scientist in the Department of Plant Pathology, Ohio State University, Ohio, USA, in 2011, working on the development of nucleic acid bases diagnostic tools for plant pathogens. He has 19 years' research experience in the area of resistance breeding, variability in plant pathogens, diagnostics, biological control of fungal pathogens and plant growth-promoting rhizobacteria. He has contributed to the development of wheat and barley varieties and has registered five genetic stocks of barley resistant to different diseases. Dr Kumar has successfully completed several externally funded research projects from different Indian government agencies. He has published more than 60 research papers and review articles in international journals of high repute and contributed many book chapters and the edited book *Microbes for Climate Resilient Agriculture* (Wiley, 2017).

Prem Lal Kashyap, PhD, received his agricultural education from Chaudhary Sarwan Kumar Himachal Pradesh Krishi Vishvavidyalaya (CSKHPKV), Palampur and Punjab Agricultural University (PAU), Ludhiana (Punjab). He is currently working as a scientist (Plant Pathology) at ICAR – Indian Institute of Wheat and Barley Research, Karnal, and has nine years of research experience. He is a dedicated researcher in the area of plant pathology, and has made an outstanding and pioneering contribution in the area of phytopathogenomics, biomarker development, biocontrol and integrated disease management. He identified five new chemical molecules for the management of wheat diseases and pests, reported five new races of yellow rust of wheat, and he has contributed in the development of "GEHOON DOCTOR" mobile app. He decoded the whole genome of several agriculturally important microorganisms, developed biomarkers for the early and precise diagnosis of plant pathogens (*Alternaria*, *Fusarium*, *Colletotrichum*, *Tilletia indica*, *Urocystis agropyri*, *Puccinia triticina*, etc.) and developed a formulation "Biogrow'" available for commercialisation by AgriInovate (ICAR) India. He has three wheat varieties to his credit and has filed two patents. Dr Kashyap has greatly contributed to resource generation through externally funded projects from DBT and other agencies of the Indian government. He also created a large pool of trained personnel in the area of plant pathology that has enabled successful product development following his approach. He has more than 100 publications to his credit, including 62 peer-reviewed research papers, 2 books, 19 book chapters, 21 popular articles and 6 technical manuals. He has visited several countries including the USA, Kenya, Mexico, Bangladesh and Bolivia for academic pursuits. For significant research contribution in the field of agricultural sciences, Dr Kashyap has been recognised with several prestigious awards including NAAS Associateship–2019, NAAS Young Scientist Award, Dr Basant Ram Young Scientist Award, Professor Abrar Mustafa Memorial Award, M.K. Patel Memorial Young Scientist Award and Professor Mahatim Singh Memorial Award.

Gyanendra Pratap Singh, PhD, works as Director of ICAR – Indian Institute of Wheat and Barley Research, Karnal, leading wheat coordinated trials and wheat research management in India. He has more than 25 years of research association with wheat variety development for temperature and moisture suppressive environments of Eastern and Central India. He also worked on the development of high-yielding wheat genotypes suitable for tillage interactions, temperature tolerance from the cropping system perspective in the rice-wheat cropping system with early and late heat tolerance. As a wheat breeder, he developed 42 wheat varieties for higher yield as well as different traits such as yellow rust resistance, short duration, suitable for rainfed and late-sown conditions and resistant to abiotic stresses. Among these, HD2967 and HD3086 are mega-varieties and occupy more than 60% of the wheat-growing area. In addition, he has registered a number of wheat genetic stocks. He has vast teaching experience and taught several courses at the graduate and post-graduate level at the Indian Agricultural Research Institute (IARI), New Delhi. He has visited several counties including the USA, Australia, Kenya and China. He has published more than 140 research papers and review articles in international journals of high repute, several book chapters and technical bulletins. For his outstanding contribution in the field of wheat genetics and varietal development, he has been awarded with prestigious national and international awards such as BGRI Gene Stewardship Award, Rafi Ahmed Kidwai Award; Fellow of the National Academy of Agricultural Sciences (NAAS), New Delhi and BP Pal Gold Medal.

Contributors

Carlos Eduardo Aucique-Pérez
Department of Phytopathology
Federal University of Viçosa
Viçosa, Brazil

Naresh Kumar Bainsla
Division of Genetics
Indian Agricultural Research Institute
New Delhi, India

Eliseu Binneck
Embrapa Soja
Londrina, Brazil

Gabriela Andriolio Camilotti
Universidade de Passo Fundo
Passo Fundo, Brazil

Cinthia Cazal
CAPECO
Asuncion, Paraguay

Alice Chavez
CAPECO
Asuncion, Paraguay

Aakash Chawade
Department of Plant Breeding
Swedish University of Agricultural Sciences
Alnarp, Sweden

Maurício Antônio de Oliveira Coelho
Epamig
Patos de Minas, Brazil

Luciano Consoli
Embrapa Trigo
Passo Fundo, Brazil

V. F. Consolo
Instituto de Investigaciones en Biodiversidad y Biotecnología (INBIOTEC-CONICET)
Fundación para Investigaciones Biológicas Aplicadas (FIBA)
Mar del Plata, Argentina

Daniel Debona
Department of Phytopathology
Federal University of Viçosa
Viçosa, Brazil

Carolina Cardoso Deuner
Universidade de Passo Fundo
Passo Fundo, Brazil

Alexei de Campos Dianese
Embrapa Cerrados
Planaltina, Brazil

Jéssica Rosset Ferreira
Universidade de Passo Fundo
Passo Fundo, Brazil

Krishna Gopal
ICAR – Indian Institute of Wheat and Barley Research
Karnal, India

Augusto César Pereira Goulart
Embrapa Agropecuária Oeste
Dourados, Brazil

Vikas Gupta
ICAR – Indian Institute of Wheat and Barley Research
Karnal, India

Caroline Hawerroth
Department of Phytopathology
Federal University of Viçosa
Viçosa, Brazil

Xinyao He
Global Wheat Program, International Maize and Wheat Improvement Center (CIMMYT)
El Batán, Mexico

Satyajit Hembram
Department of Plant Pathology
Uttar Banga Krishi Viswavidyalaya (UBKV)
Pundibari, India

Prem Lal Kashyap
ICAR – Indian Institute of Wheat and Barley Research
Karnal, India

Man Mohan Kohli
CAPECO
Asuncion, Paraguay

Sudheer Kumar
ICAR – Indian Institute of Wheat and Barley Research
Karnal, India

Sunita Mahapatra
Department of Plant Pathology
Bidhan Chandra Krishi Viswavidyalaya, Mohanpur
Nadia, India

I. Martinez
Facultad de Ciencias Agrarias y Forestales de la Universidad Nacional de La Plata (FCAyF UNLP)
Fitopatología-CIDEFI-CONICET
La Plata, Argentina

A. E. Perelló
Facultad de Ciencias Agrarias y Forestales de la Universidad Nacional de La Plata (FCAyF UNLP)
Fitopatología-CIDEFI-CONICET
La Plata, Argentina

Jonas Alberto Rios
Department of Phytopathology
Federal University of Viçosa
Viçosa, Brazil

Fabrício Ávila Rodrigues
Department of Phytopathology
Federal University of Viçosa
Viçosa, Brazil

Sandra Maria Mansur Scagliusi
Embrapa Trigo
Passo Fundo, Brazil

Claudine Dinali Santos Seixas
Embrapa Soja
Londrina, Brazil

Palika Sharma
ICAR – Indian Institute of Wheat and Barley Research
Karnal, India

Rajiv K. Sharma
Global Wheat Program, International Maize and Wheat Improvement Center (CIMMYT)
National Agricultural Science Complex (NASC)
New Delhi, India

Gyanendra Pratap Singh
ICAR – Indian Institute of Wheat and Barley Research
Karnal, India

Pawan K. Singh
International Maize and Wheat Improvement Center (CIMMYT)
El Batán, Mexico

Gisele Abigail Montan Torres
Embrapa Trigo
Passo Fundo, Brazil

1 Wheat Blast
A Global Threat to Wheat Production

Rajiv K. Sharma and Pawan K. Singh

CONTENTS

1.1 WHAT IS WHEAT BLAST?

Blast is a familiar term to pathologists, especially rice blast, which is one of the most important and much investigated diseases of rice. The blast-causing pathogen *Magnaporthe* spp. (synonym *Pyricularia* spp.) is pathogenic to more than 50 graminaceous hosts, including food security crops such as rice, wheat, finger millet, pearl millet and foxtail millet (Kharkar et al., 2016). The wide host range of this fungus speaks of the dynamic genetic-pathotypic variability emanating from the variability-generating reproductive system of the fungus (Moreira et al., 2015). Wheat blast is considered a spike disease; however, its symptoms can be observed on all aerial plant parts (Igarashi, 1990). Blast affects the leaves of highly susceptible varieties (Maciel et al., 2008; Urashima et al., 2009) and leaf symptoms include grey-green, water-soaked lesions with dark green borders. The lesions turn light tan with necrotic borders when fully expanded (Filha et al., 2011). Leaf lesions rarely precede spike blast; therefore, the importance of the inoculum originating from the leaves in severely affected fields is disputed (Cruz et al., 2015b). Since correct and uniform disease scoring is a prerequisite to comparative studies, reporting and interpreting blast scoring is another aspect that is still evolving. Maciel et al. (2013) employed the software "ImageJ" to propose a diagrammatic scale to score blast on wheat spikes. Similarly, Rios et al. (2013) developed standard area diagram sets (SADs) to quantify blast on wheat leaves. Wheat blast is a threatening disease as it can render shrunken, shrivelled and deformed grains within a week of initial symptoms. Its severity can be quite shocking to the farmer, giving them no time to react. It has the potential to cause up to 100% loss under the right disease-favouring conditions.

1.2 WHERE IS IT PRESENT?

Wheat blast was first sighted in 1985 in Parana State, Brazil. From Brazil, it soon spread to neighbouring countries including Bolivia (1996), Paraguay (2005) (Kohli et al., 2011) and Argentina (Perello et al., 2015) (Table 1.1). It can cause up to 100% yield loss and has already become a serious production constraint in the tropics and sub-tropics,

TABLE 1.1
Countries where Blast Has Been Reported and/or Pathogenicity Has Been Confirmed

No.	Country/region	Blast pathogen found on or pathogenic to the crop/species	References
1	US	Perennial ryegrass (*Lolium perenne*), wheat, triticale, rice, tall fescue *(Festuca arundinacea*), rye and 12 other species	Viji et al. (2001); Rush et al. (1972); Farman et al. (2017)
2	Bangladesh	Wheat	Sadat and Choi (2017); Islam et al. (2016)
3	Brazil	*Avena* spp., wheat, barley	Oh et al. (2002); Castroagudín et al. (2016)
4	Bolivia	Wheat	Kohli et al. (2011)
5	Paraguay	Wheat	Kohli et al. (2011)
6	Korea Republic	Rice, crabgrass (*Digitaria sanguinalis*), tall fescue (*Festuca arundinacea*), wheat, maize, rye, oat, green foxtail (*Setaria viridis*)	Choi et al. (1996)
7	India	Rice, finger millet (*Eleusine coracana*)	Todmal et al. (1994)
8	Japan	Barley, wheat, *Avena sativa*, triticale, rice	Ikegami (1973)
9	India	Rice, maize, oats	Thomas (1940)
10	Argentina	Wheat	Perello et al. (2015)

including Argentina, Brazil, Paraguay and Bolivia. The first appearance of wheat blast beyond South America was reported from Bangladesh in 2016 with the potential to spread to other major wheat-producing areas of Bangladesh, India and Pakistan because of the existing favourable conditions for the blast pathogen (Sadat & Choi, 2017). The rice-blast pathogen as such has been reported from almost all rice-growing countries.

1.3 PATHOGEN

Wheat blast is caused by *Magnaporthe oryzae* pathotype *Triticum* (MoT) (Maciel, 2016; Castroagudín et al., 2015, Perello et al., 2015; Maciel et al., 2014). Since the disease is of recent origin, researchers refer to the pathogen variously as *Pyricularia grisea* (Filha et al., 2011; Kohli et al., 2011; Rocha et al., 2014), *Pyricularia oryzae* (Araujo et al., 2016; Oliveira et al., 2015; Cruz et al., 2015a; Silva et al., 2015), *Magnaporthe grisea* (Urashima & Kato, 1994; Peng et al., 2011; Pagani et al., 2014) and *M. oryzae Triticum* (Cruz et al., 2015b). The blast pathogen has been observed to have a wide host range and is known to infect many cereals such as rice, wheat and barley, as well as grasses. The pathotypes of MoT are a subpopulation within *M. oryzae* and are different from subpopulations infecting other cereals such as rice (the *Oryzae* pathotypes, MoO), finger millet (the *Eleusine* pathotypes) and foxtail millet (the *Setaria* pathotypes). However, the wheat blast pathogen differs from the rice-blast pathogen in terms of inheritance of pathogenicity (Urashima, 1999). Oh et al. (2002) discovered a temperature-dependent virulence of an *Avena* isolate on wheat. Wheat was susceptible to this isolate at 28°C but became resistant at 20°C. This temperature-dependent resistance was found to be associated with an increased incidence of papilla formation and a hypersensitive reaction (HR). Previous to this study, Urashima and Kato (1998) in a greenhouse study had reported the role of the development stage in blast disease in cross-inoculation studies. Wheat was susceptible to blast strain from *Setaria geniculate* at seedling stage, but was resistant at the adult plant stage. In contrast, wheat was resistant to the strain from *Brachiaria plantaginea* at seedling stage, but was susceptible at the adult plant stage. The super pathogen complex of *M. oryzae–M. grisea* comprises several host-specific subgroups virulent on several species, but there is no barley-specific subgroup. To characterise the relationship between barley and various subgroups of the pathogen, Hyon et al. (2012) inoculated 24 barley cultivars with each of 18 isolates from various hosts. The cultivar–pathogen interaction ranged from non-host-like immune responses to typical host responses. At the gene level, the immune response emanated from the gene *PWT1*, whereas at the cytological level, the immune response was associated with both papilla formation and hypersensitive reaction. The authors opined that these interactions could reveal various steps in the process of host specialisation of a parasite species and the concomitant evolution of resistance. Klaubauf et al. (2014) attempted to clarify the taxonomic relationship among species that are *Magnaporthe* or *Pyricularia* like in morphology. They proposed 10 novel genera and 7 novel species, and also concluded that neither host range can be used as a taxonomic criterion without extensive pathotyping nor can the conidium morphology be used as a taxonomic criterion at the generic level without phylogenetic data.

The blast pathogen can spread through seed and can survive on crop residues. An infected rachis passes on the pathogen to the harvested seed. Goulart et al. (1995)

found that no seed from variety BH1146 carried the blast pathogen as this variety had the least infection index contrary to the variety Anahuac that had highest infection index (99.5%) with 26.7% of its seeds carrying the wheat blast pathogen. They also reported a significant positive correlation between the field incidence of wheat blast and the percentage of seed-carrying blast pathogen. A study involving the rice-blast pathogen revealed that crop residue was a more efficient source of inoculum as compared to seed (Raj & Pannu, 2017).

1.4 WHY AND HOW CAN IT BE A GLOBAL THREAT?

A disease becomes a threat if the region in question grows susceptible cultivars and the climate also favours disease development. The other major factors contributing to the spread of a disease is a pathogen that is potent, fast evolving and adapts to non-traditional areas.

1.4.1 Favourable Climatic Conditions

Wheat blast requires concurrent heat and humidity to develop. Cardoso et al. (2008) observed the highest blast intensity at 30°C, which increased with wetting duration. The lowest severity was observed at 25°C with 10 h of spike wetness. Irrespective of the temperature regime, a wetting duration of less than 10 h caused no blast, whereas a temperature regime of 25°C with a 40 h wetting duration caused the severity to exceed 85%. Similar conditions during the wheat life cycle exist in several parts of the world. Cardoso et al. (2008) developed a model that shows the blast intensity as a function of temperature and spike wetness duration. Urashima et al. (2007) studied the aerial dissemination of the blast pathogen to favourable climates. They employed molecular tools and concluded that pathogen clones could travel up to 1000 m from the inoculum source. Given that the global trade in wheat is a reality, experts had earlier warned (Peng et al., 2011) about a possible movement of blast from Latin America to similar regions of Africa and Asia. These fears came true when wheat blast was detected in Bangladesh in 2016 (Callaway, 2016). Islam et al. (2016) suspected that the wheat blast of Bangladesh was caused by a wheat-infecting South American lineage of the blast pathogen *M. oryzae*. If the pathogen shows such migratory capabilities in South Asia, it could jeopardise the food security of 300 million inhabitants of this region who consume about 100 million tonnes of wheat each year. The detection of blast in Bangladesh has already shown that the disease is capable of striking all parts of the globe from east to west wherever favourable climatic conditions exist. Even more worrying is the fact that fungicides, at best, offer only a partial defence (CIMMYT, 2019).

1.4.2 Pathogen is Evolving Fast and Evolving to Wide Spectrum Host–Pathogen Interactions and New Hosts

The blast pathogen is capable of rapidly evolving to adapt to new climates and new species. From its early years in South America, the wheat blast pathogen demonstrated that it was capable of evolving to attack new species and diverse climates.

Initially, it remained confined to Brazil and other countries of the Southern Cone of Latin America. Its origin was a matter of great speculation until it was demonstrated that the causal agent was different from the rice-blast pathogen. Conducting blast pathogen host range experiments, Todmal et al. (1994) reported that the *M. grisea* isolate from finger millet could infect barley and wheat but not maize, oats, rice, sorghum, etc. Conversely, an isolate of rice could not infect finger millet. Viji et al. (2001) also reported that the *P. grisea* affecting *Lolium* grass was similar to the one causing blast on wheat and triticale but was different from the one causing rice blast. An oats-specific isolate was reported by Oh et al. (2002). This isolate remained avirulent on wheat at 20°C; however, it caused blast on wheat when temperatures increased to 28°C. When triticale was attacked for the first time in epidemic form in 1995, Mehta and Baier (1998) investigated blast pathogenicity on triticale. They investigated eight isolates from triticale on wheat, triticale, rice, barley and rye genotypes. One isolate from rice and another from wheat were also involved. They found isolates from triticale aggressive to triticale and oats, less aggressive to wheat and not compatible with rice. The rice isolate was able to infect some wheat cultivars. They concluded triticale isolates to be of a different origin from the wheat and rice isolates. Based on sexual characteristics and DNA fingerprinting, indicating the ability to evolve, an *M. oryzae* isolate causing blast on *Lolium* spp. was reported to be unstable in genomic organisation (Tosa et al., 2004). Urashima et al. (2005) identified two distinct subpopulations of *M. grisea* causing wheat blast in Brazil. Similarly, the pathogen was discovered to have developed a wide spectrum of host gene–pathogen interaction, endowing itself with flexibility to attack new genotypes and even new species. Urashima et al. (2004) sampled 72 monoconidial isolates and inoculated 20 wheat cultivars and found 54 distinct virulence patterns. The degree of variability is evident from the fact that only 22% of the isolates showed similar virulence patterns. None of the cultivars was resistant to all the isolates and cultivars differed in their degree of resistance. A new *Pyricularia*-specific phytotoxin "pyrichalasin" was reported to be exclusively associated with the genus-specific pathogenicity of *Digitaria* isolates (Tsurushima et al., 2005). Similarly, Urashima and Silva (2011) reported a new variant of the blast pathogen when they found that *M. grisea* on black oats formed a homogeneous and genetically distinct group of its own, different from other hosts viz., rice, wheat, triticale and barley. However, they found that one of the isolates viz., isolate 15720, was similar to wheat isolate. They found no sexual compatibility between *M. grisea* from oat and other hosts; however, isolates from oat could infect triticale, wheat, rye, millet, barley, sorghum and rice; also, cross-inoculation was positive with wheat, triticale, rice and barley. Studying the phylogenetic relationships among *M. grisea* isolates from different crops, Ha et al. (2012) employed Amplified Fragment Length Polymorphism (AFLP) to differentiate between wheat- and rice-infecting strains of the blast-causing pathogen *M. grisea*. They discovered that isolates from wheat predominantly clustered with isolates from finger millet and rye grass in genetic distance analysis, suggesting an evolutionary relationship. After its detection on wheat in Brazil in 1985, it remained confined to wheat and black oats. However, it did spread to some other Gramineae species, but white oat remained a non-host until 2012 when the cultivar IAC7 fell to blast in Sao Paulo (Marangoni et al., 2013). The first visual detection of blast in the

US in 2011 was also reported to be a case of host jumping due to the disruption of a single avirulence gene in the blast pathogen (Tosa et al., 2016), indicating a dynamic genetic system capable of gaining virulence over new host species. Farman et al. (2017) concluded that the wheat blast–causing pathogen that they discovered most likely evolved from an endemic *Lolium*-infecting pathogen, and such *Lolium* isolates capable of infecting wheat had existed in the US for at least 16 years. Recently, Inoue et al. (2017) demonstrated that the wheat-infecting *Triticum* pathotype originally derived from an unknown South American *M. oryzae* strain via a loss-function mutation in the *PWT3* gene. Blast pathogen isolates from different species displayed differential infection abilities and host–parasite specificity between wheat cultivars and pathogen isolates (Peng et al., 2011). The blast pathogen isolated from triticale and barley could infect barley, wheat, triticale, oat and rye, but not rice, sorghum, maize, common millet, sugarcane or *Brachiaria brizanta* (Urishama et al., 2004). Investigations into discovering new hosts for the wheat blast pathogen revealed that weeds viz., *Bromus catharticus*, *Digitaria horizontalis* and *Lolium multiflorum* served as alternative hosts (Chavez & Kohli, 2015). Pordel et al. (2016) reported corn, banana, Bermuda grass, paspalum and *Elusine* as new hosts for *Pyricularia* in Iran. The high genetic-pathotypic variability observed in the blast-causing fungus is perhaps also responsible for the low durability of resistance and may also be driving a host shift by the pathogen. Comparative genomics analysis of host-specific *M. oryzae* isolates obtained from rice, foxtail millet, finger millet, wheat and oats and also one *M. grisea* isolate revealed gain and loss of genes as a major evolutionary mechanism driving specialisation to *Oryzae* and *Setaria* (Yoshida et al., 2016). The study also attributed transposable elements responsible for gene evolution possibly by enhancing chromosomal rearrangements and other forms of genetic variation. Moreira et al. (2015) also opined that an understanding of the regulatory mechanism of the pathogen sexual cycle and reproductive biology is essential for blast disease management.

1.5 SEXUAL REPRODUCTION EVOLVES NEW STRAINS

The blast pathogen in South America is still evolving and spreading horizontally to cover large geographical areas. Its detection for the first time in 1985 raised many questions such as where had it come from? Was it the same pathogen that caused rice blast? Urashima and Kato (1998) concluded that rice blast and wheat blast pathogens were different. They sampled isolates from several species and studied their relationship with wheat blast. The lack of cross-infection between *M. grisea* strains from wheat grown near rice and rice grown near wheat led them to conclude that wheat blast did not originate from the rice-blast pathogen. The isolate from *S. geniculate* was most similar to the wheat blast fungus as cross-inoculation was successful. Castroagudín et al. (2016) reported the detection of a new *Magnaporthe* species affecting wheat and Poaceae species; 128 isolates of *M. oryzae* were grouped into three clades. Clade 1 isolates were associated with rice, corresponding to the previously described rice-blast pathogen and were termed *P. oryzae* pathotype *Oryzae* (PoO). Clade 2 isolates were associated exclusively with wheat and were named *P. oryzae* pathotype *Triticum* (PoT). Clade 3 had isolates obtained from wheat as well as other Poaceae hosts. Clade 3 was distinct from *M. oryzae* and represented a new species, *Pyricularia graminis-tritici* (*PgT*). The study did not find any morphological difference among the species, but they displayed a distinctly different pathogenicity spectrum. However, Gladieux et al. (2018) recently refuted the existence of *PgT* via refined phylogenetic studies using whole-genome analysis of 76 fungal strains from different hosts. In their study, the *M. oryzae* isolates were resolved into numerous lineages, each infecting a limited number of host species. Additionally, the study provided evidence that interlineage gene flow contributed to the genetic makeup of multiple *M. oryzae* lineages within the same species. Bruno and Urashima (2001) studied the sexual behaviour of the blast pathogen from various hosts and reported that isolates from *B. plantaginea*, as well as from *Triticum aestivum*, possibly play a role in the variability of the wheat blast disease. Since the blast strain from *S. geniculate* possessed genetic variability, it was also suggested to cause variability in the wheat blast fungus. The rice-blast strain showed low sexual fertility. Characterising the sexual compatibility of *P. grisea* among isolates from different hosts, Galbieri and Urashima (2008) observed that isolates from wheat were sexually compatible with isolates from *B. catharticus*, *Phalaris canariensis* and *Triticosecale*. The pathogen as such has no subdivision among isolates collected from wheat fields of central-western, south western and southern Brazil, which indicates a high level of gene flow across a large geographical expanse. Maciel et al. (2014) concluded that populations of the wheat blast pathogen exhibited a mixed reproductive system in which sexual reproduction was followed by local dispersal of clones. They conducted seedling virulence assays with local wheat cultivars and found 14 pathotypes in the current population; however, detached head virulence assays could differentiate only eight virulence groups on the same set of wheat cultivars leading to the conclusion that there was no correlation between seedling and head reactions.

Preira et al. (2014) characterised 31 blast isolates with 38 genomic simple sequence repeats (SSR) loci and reported 15 polymorphic loci. Some isolates collected from different states and different years demonstrated 100% similarity, whereas the rice isolate was clearly different from the wheat isolates indicating isolates differentiated when moved from one species to another. Nhani et al. (2016) employed simple sequence repeats markers to study polymorphism among various *M. oryzae* (*Triticum* haplotype – MoT) isolates. They designed 38 SSR markers utilising the genome sequence of different MoT isolates, compared the informativeness of those markers with another 52 from the literature and mapped the polymorphic ones. From among 90 SSR markers thus employed, 53 were found to be polymorphic. Clustering separated one isolate from rice from the rest of the isolates from wheat. Clustering further revealed that the pathogen is widely dispersed across wheat-growing regions in Brazil. Shirke et al. (2016) studied the genomic basis of the host specificity of various *Magnaporthe* species. They sequenced the whole genome of the *Magnaporthe* isolates infecting rice (leaf and neck), finger millet (leaf and neck), foxtail millet (leaf) and buffel grass (leaf). They investigated the gene content variability, repeat element distribution, genes involved in carbohydrate metabolism, etc. The study revealed independent genome evolution orchestrated by multiple genomic factors. Kharkar et al. (2016) also reported the rice-blast pathogen's evolutionary adaptiveness for each hot spot as revealed by AFLP markers. They discovered 102 polymorphic alleles using 9 primer pair combinations. They also reported that the pathogen moved from one location to another as revealed by the grouping of isolates from different geographic regions in one genotype. A total of 18 isolates of the blast pathogen collected from wheat were characterised with 8 microsatellite primers and were also inoculated in the seedlings of 70 wheat genotypes (Cruz et al., 2009). The primers revealed more than 75% similarity among the isolates, whereas in the virulence analysis 15 out of 18 isolates showed more than 85% similarity. It was concluded that the absence of resistance to all isolates confirmed the need to seek new sources or better recombinants of wheat that were more resistant to the disease.

1.6 EVOLVING TO BE TOLERANT TO FUNGICIDES

The blast pathogen's evolution is not limited to only attacking newer species but it is also evolving to be resistant to fungicides currently used to manage the disease. Oliveira et al. (2015) investigated the resistance offered by two fungicides viz., azoxystrobin and pyraclostrobin, against two groups of *P. oryzae* isolates from wheat. Both of these fungicides are quinone oxidoreductase inhibitors or QoI fungicides. They observed a high level of resistance among isolates to the fungicides used and concluded this to be the result of high selection pressure exerted by consecutive years of strobilurins application for the management of wheat diseases in Brazil. Similarly, Castroagudín et al. (2015) also reported the widespread distribution of QoI resistance in *M. oryzae* isolates across central and southern Brazil. They analysed the *cytochrome b* (*cyt b*) gene and found nine haplotypes, four of them carrying the G143A mutation associated with QoI resistance. The frequency of the fungicide-resistant mutants in the wheat-infecting population increased from 36% in 2005 to 90% in 2012. These results indicate an imperative need to re-examine the use of strobilurins to manage fungal wheat diseases in Brazil.

1.7 HOW SHOULD THE DISEASE BE MANAGED?

The disease was reported in Asia in 2016 and has the potential to be a major threat to food security for the subcontinent and beyond. Currently, no single disease management strategy is good enough to keep yield losses under control; therefore, an integrated management strategy seems to be the best approach. Since it is already known that the pathogen has evolved to develop resistance to some popular hitherto used fungicides, it is even more relevant to adopt an integrated approach rather than rely on fungicides alone.

1.7.1 Quarantine

Blast has reached a stage that it has limited the expansion of wheat production in Brazil (Ceresini et al., 2019). For countries and regions that have not been affected by the disease, quarantine is no doubt the most effective way of preventing the introduction and spread of wheat blast. Blast infection in fields enters seeds

(Goulart et al., 1995), thus mandating restrictions on the movement of harvest from infected fields. Great lessons should be learned from the blast outbreak in Bangladesh. In all probability, Bangladesh imported grain that somehow reached farms and was used as seed. It needs to be ensured that grains imported from blast-infected countries reach only the flourmills and any seed import, especially those from blast-affected regions, goes through strict quarantine procedures to prevent the introduction of the blast pathogen. Mottaleb et al. (2018) estimated that just 5–10% of crop loss due to blast would cause economic losses to the tune of US$132–$264 million in Bangladesh and the neighbouring countries of India and Pakistan if the disease was not properly quarantined.

1.7.2 "Wheat Holiday"

The strategy is to suspend wheat cultivation in Bangladesh and the neighbouring regions in India for a few years to eliminate the disease from the region. This would be helpful in reducing the inoculum concentration and slowing down the blast spread. However, elimination of the disease would be very difficult, considering the wide host spectrum and the airborne nature of MoT, and the difficulty in strictly executing the policy. An ex-ante study has been done to evaluate the feasibility of this policy, and the results have shown that only a few crops would be profitable in replacing wheat in Bangladesh (Mottaleb et al., 2019)

1.7.3 Disease Forecasting

The resemblance of a blast-affected spike with Fusarium head scab is quite confusing. Yasuhara et al. (2018) developed a DNA amplification-based blast detection kit to confirm the disease. The kit has the potential for use in surveillance as well as for identifying non-wheat species that may serve as a reservoir or source of inoculum for nearby wheat fields. So far, very few wheat blast–forecasting models have been proposed. Fernandes et al. (2017) analysed historical epidemics and weather patterns in the Parana State of Brazil to propose a prediction model that correctly predicted the 2015 epidemic and the 2016 non-epidemic. The study made use of the disease and weather data sets from 2001 to 2012, which included two major epidemic years viz., 2004 and 2009. They concluded that the widespread, unusually humid and warm weather during a 60-day period preceding wheat heading was the key determinant of the major outbreak. Another Brazilian study (Cardoso et al., 2008) compared the joint effect of temperature and the spike wetness period on wheat blast under controlled environment conditions. The highest blast intensity was observed at 30°C, which increased as the duration of the spike wetness period increased. The study proposed tables that can be used to predict the intensity of wheat blast based on the temperatures and the durations of wheat spike wetness obtained in the field. To further understand the disease and perfect the forecasting techniques, more work is needed to identify the favourable environmental conditions for wheat blast, alternative hosts of MoT where it survives off-season and the way MoT spores survive off-season, spreading and infecting wheat spikes.

1.8 BREEDING FOR RESISTANCE

Host resistance is an effective, economical and environmentally friendly strategy for controlling various diseases of wheat. Efforts have been made to screen germplasm and identify sources of resistance to this disease. The most resistant sources carried the 2NS translocation (Cardozo et al., 2019), but even this resistance seems to be dependent on the genetic background in which it is housed. Cellular investigations revealed that resistance to non-adapted *Magnaporthe* isolates was due to the formation of appositions beneath pathogen penetration sites that adapted virulent isolates were able to breach (Tufan et al., 2009). They also reported differential transcription post-infection between adapted and non-adapted isolates. Several major genes based on seedling reactions for wheat blast resistance viz., *Rmg1*, *Rmg2*, *Rmg3*, *Rmg4* and *Rmg5*, have been reported (Peng et al., 2011). Tosa et al. (2016) reported the interesting case of host jumping involving the *Rmg6* gene. They discovered that the avirulence of a *Lolium* isolate on common wheat was disrupted, suggesting that a mechanism of host jumping exists in the blast pathogen. Furthering the efforts, Vu et al. (2018) reported the identification of *Rmg7* and *Rmg8*, wheat genes resistant to wheat blast fungus. Since both of these resistance genes recognise the same avirulence gene in the pathogen, these two are equivalent to a single gene from a breeding perspective. Taking a cue from the wheat rust pathogen, Cruz et al. (2010) screened 70 wheat genotypes at young stage to 18 isolates of the blast pathogen. They observed BRS229, BRS179, CNT8, BRS120 and BRS Buriti having better resistance. Twelve of these 70 genotypes were also inoculated at adult stage and genotypes such as CNT8, NE20156-B, PF844001, PF964009 and PF804002 had less leaf and head area affected by blast. In another blast-screening study across species, Urashima and Kato (1994) used 7 *Triticum* spp., 18 *Aegilops* lines and another set of 43 wheat lines against *M. grisea* inoculation under greenhouse conditions. They found only two *Aegilops* accessions resistant and all others screened were susceptible. It is interesting to note, however, that the wheat leaf rust resistance gene *Lr34* confers resistance to blast in rice (Krattinger et al., 2016). Since the resistance of 2NS against wheat blast has not been complete and is showing signs of breaking down (Singh et al., 2016), it has become imperative to search for novel sources of resistance. The International Maize and Wheat Improvement Centre (CIMMYT), in collaboration with the National Agricultural Research System (NARS) from Bangladesh and Bolivia, has undertaken large-scale, multi-location screening of synthetics, wild relatives and other germplasm sourced from South Asian countries, China, etc., for this purpose.

1.9 FUNGICIDE APPLICATION

Ever since the first report of wheat blast, efforts have been underway to search for effective chemical ingredients against MoT. Many chemicals have shown encouraging results in controlling the disease in laboratory and experimental plots, but their efficacy in large-scale commercial plots, especially during epidemic years, is yet to be ascertained. In addition to searching for effective chemicals, their application technology should also be studied (Duveiller et al., 2016). Investigating the effect of the fungicide "Opera" (fungicide 13.3% epoxiconazole + 5% pyraclostrobin) on wheat blast infection, Rios et al. (2017) found that it increased the mean yield more in the resistant variety BR-18 than in the susceptible variety Guamirim. The same study reported elsewhere (Rios et al., 2016) concluded that blast was managed more effectively when resistance and fungicides were combined with over 70% and 90% control of final incidence and severity, respectively. The blast pathogen has evolved to acquire resistance to fungicides extensively used to manage this disease (Oliveira et al., 2015). Castroagudín et al. (2015) reported similar results of widespread resistance to QoI (group of fungicides used to manage blast) in *M. oryzae* populations sampled from wheat fields and Poaceous hosts across central and southern Brazil. This resistance is a result of mutation G143A which led to the evolution of the *cytochrome b* gene. Since strobilurins are widely used to manage wheat blast in Brazil, there has been a surge in the frequency of the G143A mutation in the wheat-infecting population of *M. oryzae* from 36% in 2005 to 90% in 2012 (Castroagudín et al., 2015). Pagani et al. (2014) advocated integrating several options for the efficient management of wheat blast. In a two-year study, they found that phosphite-treated plots increased yield from 9% to 80%, silicon (Si) treatment from 26% to 92% and synthetic fungicides from 90% to 121%. Rocha et al. (2014), however, concluded that control of wheat blast by means of fungicides application was effective for flag leaves but not for ears. The positive contribution of Si in augmenting the resistance to blast was confirmed by Cruz et al. (2015a). They reported limited colonisation of +Si plants by the pathogen and that this was associated with the deposition of phenolic compounds. They also observed that expression of all defence-related genes was significantly increased on infection, but that the expression level was two to three times higher for +Si plants compared to –Si counterparts. Silva et al. (2015) reported similar results of increased Si concentration causing reduced fungal growth. They found that at the histochemical level, Si is involved in the potentiation of the biosynthetic pathway of flavonoids that increases wheat resistance to blast. In an experiment reported by Filha et al. (2011), silicon application reduced the area under blast progress curve by 31%. Several substances, such as jasmonic acid (JA), deacetylated chitosan (DC), potassium silicate (PS), potassium phosphate (PP) and tebuconazole (TE), have been experimented with to manage wheat blast (Cruz et al., 2011). Cruz et al. (2011) found that PP was the best treatment that most reduced severity on the three cultivars tested. When added to the culture medium, TE and PS gave the lowest values for mycelial growth. They concluded that PP and TE increased the potentiation of wheat resistance to blast, which was also dependent on the inherent level of resistance of the cultivar. According to Urashima and Kato (1994), probenazole and tricyclazole gave good control of blast, except at the heading stage. They also reported that new products containing blacin and acetamide gave good protection to the wheat head. Some combinations of earlier-reported fungicides viz., tricyclazole and tebuconazole were reported (Goulart & Paiva, 1993) to give the best yield increase if followed by thiophanate-methyl + mancozeb. However, Goulart et al. (1996) reported only mancozeb application to be economically viable. Even though the disease is new and is a subject of detailed investigations worldwide, it is a disease with serious consequences; therefore, early warning or disease forecasting can be of great help for farmers and administrators. The development of wheat blast requires the simultaneous occurrence of temperature and spike wetness. Fatema et al. (2018) studied the effect of several plant extracts and found garlic extract to be an effective controller of wheat blast under both *in vitro* and *in vivo* conditions.

1.10 CULTURAL PRACTICES

It is important to time the sowing date (Coelho et al., 2016) to avoid the coincidence of the heading stage with blast-favourable conditions, which varies in South America and South Asia. In the former, late planting is recommended for wheat to escape severe infection, but in the latter early planting is usually preferred. The effectiveness of rotation has not been verified, but it is assumed less effective since many cereal crops are susceptible to MoT. On the other hand, nitrogen has long been known to induce susceptibility to blast in rice (Ballini et al., 2013). The authors reported a gene, *N1S1*, which increased blast susceptibility under high nitrogen supply. However, a wheat holiday has been proposed (Mottaleb et al., 2019) as a potent strategy to curtail the spread of this disease from the Bangladesh side to the neighbouring Indian wheat area. Deep ploughing and the elimination of alternative hosts from the fields might make a positive contribution to disease control, but are labour intensive and not economical.

1.11 BIOLOGICAL CONTROL

There is an increasing interest in the biological control of fungal diseases in general and blast of rice and wheat has not remained untouched. Several anti-rice blast biocontrol agents (BCA) have been identified and are being investigated (Yu et al., 2013; Singh et al., 2012; Shimoi et al., 2010). Cabot et al. (2018) studied the role of two BCAs viz., *Pseudomonas stutzeri* AN10 and *Trichoderma harzianum* Th56, in salt-stressed seedlings infected with two *P. oryzae* isolates. They found that BCA improved the performance of salt-stressed wheat plants infected with a less aggressive blast pathogen. The lower infection in salt-stressed BCA-treated plants could be due to a salt-induced priming state required to trigger an early expression of plant defence genes.

1.12 CONCLUSION AND FUTURE OUTLOOK

The wheat researchers' community in the Indian subcontinent was busy improving yield and managing fewer major yield constraints such as rusts, bunts, blights and blotches, when it was suddenly shaken by the discovery of blast on wheat in Bangladesh in 2016. It was in a farmer's field and perhaps there was no seed import from any blast-infested country, but definitely a grain shipment (Islam et al., 2016) found its way into the farmer's fields as seed. Although a new disease to this part of the world, researchers (Peng et al., 2011) had been warning of the potential threat. It is not a very old disease and a full understanding of its epidemiology, reproduction, variability patterns, etc., is still lacking (Ceresini et al., 2019). Since its pathogen is highly variable, it is evolving fungicide-resistant strains (Oliveira et al., 2015; Castroagudín et al., 2015), its host resistance is far from complete (Singh et al., 2016) and effective fungicides and their application methodologies are still being identified (Cruz & Valent, 2017), hence an integrated disease management strategy is the way forward. There is a need to develop appropriate agronomic interventions (Coelho et al., 2016) and biocontrol agents (Cabot et al., 2018), which have been found to enhance the efficacy of fungicides and host resistance (Rios et al., 2016). Although some work has been undertaken to develop forecasting systems, there is a need to have similar information for local geographies. The disease has the potential to assume very serious dimensions and about seven million wheat crops are at risk (Mottaleb et al., 2018). There is a need to employ a multipronged strategy and launch an all-out coordinated effort to manage this menace.

REFERENCES

Araujo, L., J. M. Soares, M. C. C. de Filippi and F. A. Rodrigues. "Cytological Aspects of Incompatible and Compatible Interactions Between Rice, Wheat and the Blast Pathogen *Pyricularia oryzae*." *Scientia Agricola* 73 (2016): 177–83.

Ballini, E., T. T. N. Thuy and J. B. Morel. "Diversity and Genetics of Nitrogen-Induced Susceptibility to the Blast Fungus in Rice and Wheat." *Rice* 6 (2013): 32.

Bruno, A. C. and A. S. Urashima. "Sexual Relationship Between *Magnaporthe grisea* from Wheat and from Other Hosts." *Fitopatologia Brasileira* 26 (2001): 21–6.

Cabot, C., R. Bosch, S. Martos, C. Poschenrieder and A. Perello. "Salinity is a Prevailing Factor for Amelioration of Wheat Blast by Biocontrol Agents." *Biological Control* 125 (2018): 81–9.

Callaway, E. "Devastating Wheat Fungus Appears in Asia for the First Time." *Nature* 532 (2016): 421–2.

Cardoso, C. A. de A., E. M. Reis and E. N. Moreira. "Development of a Warning System for Wheat Blast Caused by *Pyricularia grisea*." *Summa Phytopathologica* 34 (2008): 216–21.

Cardozo, T. L., A. Chavez, N. Bobadilla, E. P. Perez and M. Kohli. "Variable Resistance of Bread Wheat (*Triticum aestivum*) Lines Carrying 2NS/2AS Translocation to Wheat Blast." *Plant Breeding* 138, no. 1 (2019): 62–8.

Castroagudín, V. L., P. C. Ceresini, S. C. de Oliveira et al. "Resistance to QoI Fungicides is Widespread in Brazilian Populations of the Wheat Blast Pathogen *Magnaporthe oryzae*." *Phytopathology* 105 (2015): 284–94.

Castroagudín, V. L., S. I. Moreira, D. A. S. Preira et al. "*Pyricularia-graminis-tritici*, a New Pyricularia Species Causing Wheat Blast." *Persoonia* 37 (2016): 199–216.

Ceresini, P. C., V. L. Castroagudin, F. A. Rodrigues et al. "Wheat Blast: From its Origin in South America to its Emergence as a Global Threat." *Molecular Plant Pathology* 20, no. 2 (2019): 155–72.

Chavez, A. and M. Kohli. "Alternative Hosts of *Magnaporthe grisea* of Wheat in Paraguay." *Revista Investigación Agraria* 17 (2015): 54–9.

Choi, W., S. Chun and Y. Lee. "Host Range of Korean Isolates of *Magnaporthe grisea*." *Korean Journal of Plant Pathology* 12, no. 4 (1996): 453–4.

CIMMYT. 2019. Assessing the effectiveness of a "wheat holiday" for preventing blast in the lower Gangetic plains. https://www.cimmyt.org/news/assessing-the-effectiveness-of-a-wheat-holiday-for-preventing-blast-by-examining-alternative-crops-in-the-lower-gangetic-plains/

Coelho, M. A. de O., G. A. M. Torres, P. R. Cecon and F. M. Santana. "Sowing Date Reduces the Incidence of Wheat Blast Disease." *Pesquisa Agropecuaria Brasileira* 51, no. 5 (2016): 631–7.

Cruz, C. D. and B. Valent. "Wheat Blast Disease: Danger on the Move." *Tropical Plant Pathology* 42 (2017): 210–22.

Cruz, M. F. A., J. L. N. Maciel, A. M. Prestes and L. Consoli. "Molecular Pattern and Virulence of *Pyricularia grisea* Isolates from Wheat." *Tropical Plant Pathology* 34 (2009): 393–401.

Cruz, M. F. A., A. M. Prested, J. L. Maciel and P. L. Scheeren. "Partial Resistance to Blast on Common and Synthetic Wheat Genotypes in Seedling and in Adult Plant Growth Stages." *Tropical Plant Pathology* 35 (2010): 24–31.

Cruz, M. F. A., A. P. C. Diniz, F. A. Rodrigues and E. G. Barros. "Aplicação foliar de produtos na redução da severidade da brusone do trigo. [Foliar Application of Products on the Reduction of Wheat Blast Severity]." *Tropical Plant Pathology* 36 (2011): 424–8.

Cruz, M. F. A., D. Debona, J. A. Rios, E. G. Barros and F. A. Rodrigues. "Potentiation of Defence Related Gene Expression by Silicon Increases Wheat Resistance to Leaf Blast." *Tropical Plant Pathology* 40 (2015a): 394–400.

Cruz, C. D., J. Kiyuna, W. W. Bockus, T. C. Todd, J. P. Stack and B. Valent. "*Magnaporthe oryzae* Conidia on Basal Wheat Leaves as a Potential Source of Wheat Blast Inoculum." *Plant Pathology* 64 (2015b): 1491–8.

Duveiller, E., X. He and P. K. Singh. "Wheat Blast: An Emerging Disease in South America Potentially Threatening Wheat Production." In *The World Wheat Book* Vol 3, edited by A Bonjean and M van Ginkel, 1107–22. Paris: Lavoisier, 2016.

Farman, M., G. Peterson, Li Chen et al. "The *Lolium* Pathotype of *Magnaporthe oryzae* Recovered from a Single Blasted Wheat Plant in the United States." *Plant Disease* 101 (2017): 684–92.

Fatema, T. Z., M. A. Kashem, M. A. Hasan and M. A. Hossain. "Effects of Plant Extracts on Controlling Wheat Blast Disease Caused by *Magnaporthe oryzae* pathotype *Triticum* in Bangladesh." *Fundamental and Applied Agriculture* 3, no. 2 (2018): 422–33.

Fernandes, J. M. C., M. Nicolau, W. Pavan et al. "A Weather Based Model for Predicting Early Season Inoculum Build Up and Spike Infection by the What Blast Pathogen." *Tropical Plant Pathology* 42, no. 3 (2017): 230–7.

Filha, M. S. X., F. A. Rodrigues, G. P. Domiciano, H. V. Oliveira, P. R. Silveira and W. R. Moreira. "Wheat Resistance to Leaf Blast Mediated by Silicon." *Australian Plant Pathology* 40 (2011): 28–38.

Galbieri, R. and A. S. Urashima. "Sexual Characterization, Compatibility and Occurrence of Sexual Reproduction Among Isolates of *Pyricularia grisea* from Different Hosts." *Summa Phytopathologica* 34 (2008): 22–8.

Gladieux, P., B. Condon, S. Ravel et al. "Gene Flow Between Divergent Cereal- and Grass-Specific Lineages of the Rice Blast Fungus *Magnaporthe oryzae*." *mBio* 9, no. 1 (2018). doi:10.1128/mBio.01219-17.

Goulart, A. C. P. and F. A. Paiva. "Evaluation of Fungicides in the Control of Wheat (*Triticum aestivum*) Blast (*Pyricularia oryzae*)." *Fitopatologia Brasileira* 18 (1993): 167–73.

Goulart, A. C. P., F. A. Paiva and P. J. M. Andrade. "Relationship Between Incidence of Blast in Wheat Heads and the Presence of *Pyricularia grisea* in the Harvested Seeds." *Fitopatologia Brasileira* 20 (1995): 184–9.

Goulart, A. C. P., F. A. Paiva, F. G. A. de Melo and A. Richett. "Effect of Timing and Number of Applications of Fungicides Tebuconazole and Mancozeb on the Control of Wheat Blast Disease (*Pyricularia grisea*): Economical and Technical Viability." *Fitopatologia Brasileira* 21 (1996): 381–7.

Ha, X., T. Wei and A. von Tiedmann. "Epidemiological and Phytopathological Studies on Wheat Blast (*Magnaporthe grisea*) – Characterization of Pathotypes, Host Specificity and Resistance in Wheat." *Julius-Kuhn-Archiv* 438 (2012): 412–3.

Hyon, G. S., T. T. N. Ngyuen, N. Thanh et al. "Characterization of Interactions Between Barley and Various Host-Specific Subgroups of *Magnaporthe oryzae* and *M. grisea*." *Journal of General Plant Pathology* 78 (2012): 237–46.

Igarashi, S. "Update on Wheat Blast (*Pyricularia oryzae*) in Brazil." In *Wheat for Non-Traditional Warm Areas. A Proceeding of the International Conference*, edited by D. A. Saunders, 480–5. Foz do Iguacu, Brazil: CIMMYT, 1990.
Ikegami, T. "On the Boundary Behaviour of Harmonic Maps." *Osaka Journal of Mathematics* 10, no. 3 (1973): 641–53.
Inoue, Y., T. P. Vy Trinh, K. Yoshida et al. "Evolution of the Wheat Blast Fungus Through Functional Losses in a Host Specificity Determinant." *Science* 357, no. 6346 (2017): 80–3.
Islam, M. T., C. Daniel, P. Gladieux et al. "Emergence of Wheat Blast in Bangladesh Was Caused by a South American lineage of *Magnaporthe oryzae*." *BMC Biology* 14 (2016): 84–94.
Kharkar, H., K. Kale and P. Thakare. "Molecular Characterization of *Pyricularia grisea* Isolates Using AFLP DNA Fingerprinting." *Journal of Pure and Applied Microbiology* 10 (2016): 1181–8.
Klaubauf, S., D. Tharreau, E. Fournier et al. "Resolving the Polyphylectic Nature of *Pyricularia* (Pyriculariaceae)." *Studies in Mycology* 79 (2014): 85–120.
Kohli, M. M., Y. R. Mehta, E. Guzman, L. de Viedma and L. E. Cubilla. "*Pyricularia* Blast – A Threat to Wheat Cultivation." *Czech Journal of Genetics and Plant Breeding* 47S (2011): 130-S–134.
Krattinger, S. G., J. Sucher, L. L. Selter et al. "The Wheat Durable, Multipathogen Resistance Gene Lr34 Confers Partial Blast Resistance in Rice." *Plant Biotechnology Journal* 14, no. 5 (2016): 1261–8.
Maciel, J. L. N. "*Magnaporthe oryzae*, the Blast Pathogen: Current Status and Options for its Control." *CABI Reviews Perspectives in Agriculture, Veterinary Science, Nutrition and Natural Resources* 6, no. 50 (2016): 1–8.
Maciel, J. L. N., A. L. D. Danelli, C. Boaretto and C. A. Forcelini. "Diagrammatic Scale for the Assessment of Blast on Wheat Spikes." *Summa Phytopathologica* 39 (2013): 162–6.
Maciel, J. L. N., E. A. Paludo, M. Só e Silva, P. L. Scheeren and E. Caierão. *Reação à brusone de genótipos de trigo do programa de melhoramento da Embrapa Trigo no estádio de planta adulta*. Passo Fundo, RS, Brazil: Embrapa Trigo, 2008.
Maciel, J. L. N., P. C. Ceresini, V. L. Castroagudín et al. "Population Structure and Pathotype Diversity of the Wheat Blast Pathogen *Magnaporthe oryzae* 25 Years After its Emergence in Brazil." *Phytopathology* 104 (2014): 95–107.
Marangoni, M. S., M. P. Nunes, N. Fonseca Junior and Y. R. Mehta. "Pyricularia Blast on White Oats: A New Threat to Wheat Cultivation." *Tropical Plant Pathology* 38 (2013): 198–202.
Mehta, Y. R. and A. Baier. "Pathogenic Variation Among Isolates of *Magnaporthe grisea* Attacking Triticale and Wheat in the State of Parana." *Summa Phytopathologica* 24 (1998): 119–25.
Moreira, S. I., P. C. Ceresini and F. Alves. "Sexual Reproduction in *Pyricularia oryzae*." *Summa Phytopathologica* 41 (2015): 175–82.
Mottaleb, K. A., P. K. Singh, K. Sonder, G. Kruseman and O. Erenstein. "Averting Wheat Blast by Implementing a 'Wheat Holiday': In Search of Alternative Crops in West Bengal, India." *PLoS ONE* 14, no. 2 (2019): e0211410 ref.63.
Mottaleb, K. A., P. K. Singh, K. Sonder et al. "Threat of Wheat Blast to South Asia's Food Security: An Ex-Ante Analysis." *PLoS ONE* 13, no. 5 (2018): e0197555.
Nhani, J. A., J. F. Preira, J. R. Ferreira, A. L. V. Bonato and J. L. N. Maciel. "Mapping Highly Informative SSR Markers in the Genome of *Magnaporthe oryzae* from Wheat." *Tropical Plant Pathology* 41 (2016): 331–5.
Oh, H. S., Y. Tosa, N. Takabayashi et al. "Characterization of an *Avena* Isolate of *Magnaporthe grisea* and Identification of a Locus Conditioning its Specificity on Oats." *Canadian Journal of Botany* 80 (2002): 1088–95.

Oliveira, S. C. de, V. L. Castroagudín, J. L. N. Maciel, D. A. Pereira, S. Dos and P. C. Ceresini. "Cross-Resistance to QoI Fungicides Azoxystrobin and Pyraclostrobin in the Wheat Blast Pathogen *Pyricularia oryzae* in Brazil." *Summa Phytopathologica* 41 (2015): 298–304.

Pagani, A. P. S., A. C. Dianese and A. C. Café-Filho. "Management of Wheat Blast with Synthetic Fungicides, Partial Resistance and Silicate and Phosphite Minerals." *Phytoparasitica* 42 (2014): 609–17.

Peng, J. L., Y. L. Zhou and Z. H. He. "Global Warning Against the Spread of Wheat Blast." *Journal of Triticeae Crops* 31 (2011): 989–93.

Perello, A., I. Martinez and M. Molina. "First Report of Virulence and Effects of *Magnaporthe oryzae* Isolates Causing Wheat Blast in Argentina." *Plant Disease* 99 (2015): 1177–8.

Pordel, A., M. Javan-Nikkhah and S. A. Khodaparast. "Revision of *Pyricularia oryzae* and Occurrence of New Hosts for the Pathogen in Iran." *Iranian Journal of Plant Pathology* 52 (2016): Pe67–83.

Preira, J. F., L. Consoli, E. A.de Bombonatto, A. L. V. Bonato and J. L. N. Maciel. "Development of Genomic SSR Markers and Molecular Characterization of *Magnaporthe oryzae* Isolates from Wheat in Brazil." *Biochemical Genetics* 52 (2014): 52–70.

Raj, R. and D. P. S. Pannu. "Host Range and Survival Studies of *Pyricularia grisea* Causing Blast Disease of Rice." *Journal of Mycology and Plant Pathology* 47 (2017): 213–20.

Rios, J. A., D. Debona, H. S. S. Duarte and F. A. Rodrigues. "Development and Validation of a Standard Area Diagram Set to Assess Blast Severity on Wheat Leaves." *European Journal of Plant Pathology* 136 (2013): 603–11.

Rios, J. A., V. S. Rios, P. A. Paul, M. A. Souza, L. Araujo and F. A. Rodrigues. "Fungicide and Cultivar Effects on the Development and Temporal Progress of Wheat Blast under Field Conditions." *Crop Protection* 89 (2016): 152–60.

Rios, J. A., V. S. Rios, P. A. Paul, M. A. Souza, L. B. M. C. Neto and F. A. Rodrigues. "Effects of Blast on Components of Wheat Physiology and Grain Yield as Influenced by Fungicide Treatment and Host Resistance." *Plant Pathology* 66, no. 6 (2017): 877–89.

Rocha, J. R. A. S. C., A. J. B. Pimentel, G. Ribeiro and M. A. Souza. "Efficiency of Fungicides in Wheat Blast Control." *Summa Phytopathologica* 40 (2014): 347–52.

Rush, M. C., G. D. Lindberg and R. B. Carver. "Blast: A Serious New Disease of Forage Grasses in Louisiana." *Phytopathology* 62 (1972): 806.

Sadat, M. A. and J. Choi. "Wheat Blast: A New Fungal Inhabitant to Bangladesh Threatening World Wheat Production." *Plant Pathology Journal* 33 (2017): 103–8.

Shimoi, S., K. Inoue, H. Kitagawa et al. "Biological Control of Rice Blast Disease by Employing Detachment Action with Gelatinolytic Bacteria." *Biological Control* 55, no. 2 (2010): 85–91.

Shirke, M. D., H. B. Mahesh and G. Malali. "Genome Wide Comparison of *Magnaporthe* Species Reveals a Host Specific Pattern of Secretory Proteins and Transposable Elements." *PLOS ONE* 11, no. 9 (2016): e0162458.

Silva, W. L., M. F. A. Cruz, A. A. Fortunato and F. A. Rodrigues. "Histochemical Aspects of Wheat Resistance to Leaf Blast Mediated by Silicon." *Scientia Agricola* 72 (2015): 322–7.

Singh, P. K., A. K. Singh, H. B. Gingh and B. K. Dhakad. "Biological Control of Rice Blast Disease with *Trichoderma harzianum* in Direct Seeded Rice Under Medium Low Land Rainfed Conditions." *Environment & Ecology* 30, no. 3B (2012): 834–7.

Singh, R. P., P. K. Singh, J. Rutkoski et al. "Disease Impact on Wheat Yield Potential and Prospects of Genetic Control." *Annual Review of Phytopathology* 54, no. 1 (2016): 303–22.

Thomas, K. M. "Detailed Administration Report of the Government Mycologist, Madras for the Year 1939–40." Department of Agriculture, Madras, 18, 1940.

Todmal, A. K., D. R. Pawar and M. S. Joshi. "Host Reactions to Fingermillet Blast (*Pyricularia grisea* Sacc.)." *Mysore Journal of Agricultural Sciences* 28 (1994): 45–6.

Tosa, Y., K. Hirata, H. Tamba et al. "Genetic Constitution and Pathogenicity of *Lolium* Isolates of *Magnaporthe oryzae* in Comparison with Host Species-Specific Pathotypes of the Blast Fungus." *Phytopathology* 94 (2004): 454–62.

Tosa, Y., Y. Inoue, T. P. V. Trinh and I. Chuma. "Genetic and Molecular Analysis of the Incompatibility Between *Lolium* Isolates of *Pyricularia oryzae* and Wheat." *Physiological and Molecular Plant Pathology* 95 (2016): 84–6.

Tsurushima, T., D. D. Le, K. Kawashima et al. "Pyrichalasin H Production and Pathogenicity of *Digitaria*-Specific Isolates of *Pyricularia grisea*." *Molecular Plant Pathology* 6 (2005): 605–13.

Tufan, H. A., G. R. D. McGrann, A. Magusin, J. B. Morel, L. Miche and L. A. Boyd. "Wheat Blast: Histopathology and Transcriptome Reprogramming in Response to Adapted and Non-Adapted Magnaporthe Isolates. *New Phytologist* 184 (2009): 473–84.

Urashima, A. S. "Genetic Analysis of *Magnaporthe grisea* Pathogenicity on Wheat." *Fitopatologia Brasileira* 24 (1999): 567–9.

Urshima, A. S., R. Galbieri and A. Stabili. "DNA Fingerprinting and Sexual Characterization Revealed Two Distinct Populations of *Magnaporthe grisea* in Wheat Blast from Brazil." *Czech Journal of Genetics & Plant Breeding* 41 (2005): 238–45.

Urashima, A. S. and H. Kato. "Varietal Resistance and Chemical Control of Wheat Blast Fungus." *Summa Phytopathologica* 20 (1994): 107–12.

Urashima, A. S. and H. Kato. "Pathogenic Relationship Between Isolates of *Pyricularia grisea* of Wheat and Other Hosts at Different Host Developmental Stages." *Fitopatologia Braileira* 23 (1998): 30–5.

Urashima, A. S., C. Grosso, A. Stabili et al. "Effect of *Magnaporthe grisea* on Seed Germination, Yield and Quality of Wheat." In *Advances in Genetics, Genomics and Control of Rice Blast Disease*, edited by X. Wang and B. Valent, 267–77. New York: Springer, 2009.

Urashima, A. S., N. A. Lavorent, A. C. P. Goulart and Y. R. Mehta. "Resistance Spectra of Wheat Cultivars and Virulence Diversity of *Magnaporthe grisea* Isolates in Brazil." *Fitopatologia Braileira* 29 (2004): 511–8.

Urashima, A. S., S. F. Leite and R. Galbieri. "Efficiency of Aerial Dissemination of *Pyricularia grisea*." *Summa Phytopathologica* 33 (2007): 275–9.

Urishama, A. S., T. D. Martins, C. R. N. C. Bueno, D. B. Favaro, M. A. Arruda and Y. R. Mehta. "Triticale and Barley; New Hosts of *Magnaporthe grisea* in Sao Paulo, Brazil: Relationship with Blast of Rice and Wheat." In *Rice Blast: Interaction with Rice and Control. Proceedings of the 3rd International Rice Blast Conference*, edited by S. Kawasaki, 251–60. September 11–14, 2002, Tsukuba Science City, Ibaraki, Japan, 2004.

Urashima, A. S. and C. P. Silva. "Characterization of *Magnaporthe grisea* (*Pyricularia grisea*) from Black Oat in Brazil." *Journal of Phytopathology* 159 (2011): 789–95.

Viji, G., B. Wu, S. Kang et al. "*Pyricularia grisea* Causing Grey Leaf Spot of Perennial Ryegrass Turf: Population Structure and Host Specificity." *Plant Disease* 85 (2001): 817–26.

Vu, L. A., Y. Inoue, S. Asuke et al. "*Rmg8* and *Rmg7*, Wheat Genes for Resistance to the Wheat Blast Fungus, Recognize the Same Avirulence Gene *AVR-Rmg8*." *Molecular Plant Pathology* 19, no. 5 (2018): 1252–6.

Yasuhara, B. J., K. F. Pedley, M. Farman, B. Valent and J. P. Stack. "Specific Detection of the Wheat Blast Pathogen (*Magnaporthe oryzae Triticum*) by Loop-Mediated Isothermal Amplification." *Plant Diseases* 102, no. 12 (2018): 2550–9.

Yoshida, K., D. G. O. Saunders, C. Mitsuoka et al. "Host Specialization of the Blast Fungus *Magnaporthe oryzae* is Associated with Dynamic Gain and Loss of Genes Linked to Transposable Elements." *BMC Genomics* 17 (2016): 370.

Yu, Q., Z. Liu, D. Lin, W. Zhang, Q. Sun, J. Zhu and M. Lin. "Characterization and Evaluation of *Staphylococcus sp.* Strain LZ16 for the Biological Control of Rice Blast Caused by *Magnaporthe oryzae*." *Biological Control* 65, no. 3 (2013): 338–47.

2 Population Biology of Wheat Blast Pathogen

Prem Lal Kashyap, Sudheer Kumar, Palika Sharma, Sunita Mahapatra, Satyajit Hembram and Gyanendra Pratap Singh

CONTENTS

2.1 INTRODUCTION

Blast caused by *Magnaporthe oryzae* (syn. *Pyricularia oryzae*) pathotype *Triticum* (MoT) is a devastating threat to wheat production (Rocha et al., 2019; Singh, 2017; Mottaleb et al., 2018; Ceresini et al., 2018; Malakar et al., 2016; Callaway, 2016; Mahmud, 2019). The wheat blast (WB) pathogen belongs to the *M. oryzae* (=*P. oryzae*) species complex and contains several host-specific subgroups. The wheat blast fungus is genetically distinct in pathogenicity from the *M. oryzae* species associated with rice blast disease (Kato et al., 2000). This pathogen has high genetic and phenotypic diversity and a broad host range, enabling it to move back and forth between wheat and other grass hosts. It has been reported that MoT develops disease lesions on wheat, *Triticum* spp. and *Aegilop* spp., while the *Oryzae* pathotype of *M. oryzae* produces disease symptoms on rice and other grasses. There is no report of host jump of *M. oryzae* from rice to wheat or vice versa (Maciel et al., 2014). Phylogenomic analysis revealed that the wheat blast isolates that caused a severe outbreak in 2016 in Bangladesh were closely related to the wheat-infecting isolates from Brazil (PY0925 and BR32) (Islam et al., 2016). Another analysis performed by Gladieux et al. (2018) suggested that WB fungus is closer to the Bolivian strain B71 than the Brazilian isolate PY0925. A recent sequence analysis has revealed that both the Bolivian and Bangladeshi isolates have a similar genotypic structure (Atc type or allele) in the avirulence gene, *PWT3*, where multiple transposable elements have been inserted (Inoue et al., 2017). Wang and colleagues (2018) reported that

TABLE 2.1
Chronological List of Significant Research Findings Made by Researchers for Defining and Understanding Wheat Blast Population Biology

Year	Research findings	References
1993	Non-rice origin of the wheat blast pathogen	Urashima et al. (1993)
2004	Existence of pathogenic races of *M. oryzae* pathotype *Triticum* in wheat	Urashima et al. (2004)
2005	Defined two distinct populations causing wheat blast epidemics in Brazil in 1998, i.e. one sexually fertile (Mato Grosso do Sul State) and another infertile (Paraná State isolates)	Urashima et al. (2005)
2008	To facilitate genetic exchange among the *Pyricularia* populations infecting different hosts, the requirement of physical proximity between wheat and other grass species under natural field conditions was proposed	Stukenbrock and McDonald (2008)
2009	Sequence comparison using a Homologue of Retrotransposon MGR583 revealed wheat-infecting MoT isolates constitute a separate, single lineage of their own, suggesting that they are derived from a single origin	Urashima et al. (2009)
2014	No subdivision among WB isolates collected from wheat fields of central-western, southwestern and southern Brazil indicating a high level of gene flow across a large geographic expanse. Moreover, populations of the wheat blast pathogen exhibited a mixed reproductive system in which sexual reproduction is followed by local dispersal of clones	Maciel et al. (2014)
2015	MoT isolates caused significantly more disease on the oldest wheat leaves compared to the youngest leaves	Cruz et al. (2015)
	Emergence of wheat blast in Brazil through a series of host shifts from populations of *Pygt* infecting native or invasive grass species growing near wheat fields	Castroagudín et al. (2015)
2016	Phylogenetic analysis based on 859,067 single nucleotide polymorphisms (SNPs) distributed across entire genomes have indicated that *Pyricularia* isolates associated with *Triticum* and *Avena* form a single monophyletic clade clearly distinct from the *P. oryzae* clade	Yoshida et al. (2016)
	Wheat blast isolates that caused the severe outbreak in 2016 were closely related to the wheat-infecting isolates from Brazil (PY0925 and BR32)	Islam et al. (2016)
2017	*Pygt* populations found on wheat and other hosts have significantly higher genetic variation than the populations of *P. oryzae* causing rice blast in Brazil	Castroagudín et al. (2017)
	Both Bolivian and Bangladeshi isolates have a similar genotypic structure (Atc type or allele) in the avirulence gene, *PWT3*, where multiple transposable elements have been inserted	Inoue et al. (2017)
	Mutations in the avirulence genes *PWT3* and *PWT4* are the only genetic factors associated with the adaptation of *Pygt* to its host, allowing the fungus to escape recognition	Inoue et al. (2017)

(*Continued*)

TABLE 2.1 (CONTINUED)
Chronological List of Significant Research Findings Made by Researchers for Defining and Understanding Wheat Blast Population Biology

Year	Research findings	References
	Wheat blast isolates that caused the severe outbreak in 2016 were closer to the Bolivian strain B71 than the Brazilian isolate PY0925	Gladieux et al. (2018)
2018	Provided evidence for historic introgression events among *Pygt* and *P. oryzae* populations	Gladieux et al. (2018)
	Avr-Rmg8 gene existed in 15 selected Brazilian isolates and Bangladeshi isolate (no. 12), suggesting that the introduction of *Rmg8* might be effective among most MoT strains	Wang et al. (2018)

the *Avr-Rmg8* gene existed in fifteen Brazilian isolates and one Bangladeshi isolate no. 12. These evidences clearly highlighted that the pathogens did not evolve from a host plant in the Asian region, but came from South American countries via the wheat trade (Islam et al., 2019; Ceresini et al., 2018).

Generally, the term "population biology" is used to describe the ecological and evolutionary dynamics of plant and pathogen populations and their interactions. The knowledge of a pathogen population structure and host–pathogen interaction is essential to understand the disease epidemiology and to devise resistance breeding and management strategies (Kashyap et al., 2019). Many factors lead to variation in a pathogen population such as migration, natural selection and recombination, which result in different populations in spatial and temporal scale across field conditions. The experimental and analytical methods commonly used to decipher the population genetics of WB fungus are depicted in Table 2.1. Various population genetic analysis and comparative genomics analysis have provided new insight into genome evolution, speciation and the origin of the pathogenicity traits of WB fungus (Ceresini et al., 2018; Gladieux et al., 2018). A series of other analytical tools have been employed to explore genomic and proteomic data to characterise patterns of mutation and natural selection. These include approaches to quantifying adaptive evolution and purifying selection as well as coalescence models to infer genome evolution, recombination patterns and effective population size (Luikart et al., 2003). In this chapter, efforts have been made to summarise past and current knowledge regarding the population biology of the wheat blast fungus and highlight researchable gaps required for a complete understanding of its ecology and population biology for effective and timely management of the wheat blast disease.

2.2 WHEAT BLAST PATHOGEN: *PYRICULARIA GRAMINIS TRITICUM (PYGT)* OR *MAGNAPORTHE ORYZAE* (=*PYRICULARIA ORYZAE*) PATHOTYPE TRITICUM (MoT)?

The taxonomy of the wheat blast–causing fungus has been disputed for many years and confusion exists between the names of *Pyricularia* and *Magnaporthe* morphs

(Ou, 1985; Zhang et al., 2016). These taxonomy disagreements have resulted in naming the wheat blast pathogen *Pyricularia grisea* pathotype *Triticum* (*Pygt*) and *M. oryzae* pathotype *Triticum* (Goulart et al., 1990; Islam et al., 2016; Valent et al., 2016). After the first report, the blast pathogen name continued changing, for example, *P. oryzae* (Oliveira et al., 2015; Cruz et al., 2015; Silva et al., 2015; Araujo et al., 2016), *P. grisea* (Filha et al., 2011; Kohli et al., 2011; Rocha et al., 2014), *Magnaporthe grisea* (Urashima & Kato, 1994; Peng et al., 2011; Pagani et al., 2014) and *M. oryzae Triticum* (Cruz et al., 2015). Castroagudín et al.'s (2016) recent designation of *Pygt* as a separate species has been based on sequence comparisons of ten housekeeping genes (actin [*ACT*], putative vacuolar import and degradation protein [*BAC6*], beta-tubulin [*βT-1*], calmodulin [*CAL*], hypothetical protein [*CH7-BAC7*], anonymous sequence [*CH7-BAC9*], chitin synthase 1 [*CHS1*], translation elongation factor 1-alpha [*EF-1α*], hydrophobin [*MPG1*] and nitrogen regulatory protein 1 [*NUT1*]) from 79 Brazilian isolates from wheat, 26 isolates from grass species obtained from wheat fields and 23 Brazilian rice isolates. They designated 18 of the aggressive wheat isolates including 8 grass hosts from Brazil as *Pyg*t and the remaining isolates were defined as *P. oryzae Triticum* (PoT) and *Oryza* pathotypes (PoO). Moreover, they also designated WB isolates of Bangladeshi origin as *Pygt*. However, a phylogenomic study conducted by Gladieux et al. (2018) systematically discarded the authenticity of *Pygt* as the name of the wheat blast fungus based on the following assumptions:

1. Accepted standards for genealogical concordance for phylogenetic sequence recognition (GCPSR) mentioned by Castroagudín et al. (2016) for defining a new species are not within the acceptable limits because only the *MPG1* gene out of 10 studied genes supported *Pygt* grouping (Gladieux et al., 2018; Valent et al., 2019).
2. During genome sequences analysis and comparison of WB isolates, Castroagudín et al. (2016) selected their own global set of WB isolates of Brazilian origin with a limited fraction of the aggressive *Triticum* lineage and the majority of isolates representing *Lolium* and *Eleusine* lineages. As a result, they obtained a *Pygt* pattern based on a single *MPG1* gene clustering and hence did not meet the GCPSR criteria recommended for defining a new species. Further, a reanalysis of the research findings of Gladieux et al. (2018) revealed that putative *Pygt* strains categorised by the MPG1 allele were discontinuously distributed within all trees built using a single nucleotide polymorphisms (SNPs)-based comparative analysis of 2682 orthologous gene sets, or with whole-genome SNP data.

Recently, Valent et al. (2019) published a similar opinion in the journal *Molecular Plant Pathology*, that the use of *Pyricularia graminis-tritici* is not an appropriate species name for the wheat blast pathogen. They objected to the statement "*Pygt is not a wheat-specialized pathogen*", published by Ceresini et al. (2019) in the same journal, and highlighted that it is logically incorrect in light of the flaws mentioned by Gladieux et al. (2018) in respect of the use of gene genealogy with a defined evolutionary history as it does not reveal the overall evolution of the species. In response to the points highlighted by Gladieux et al. (2018), Ceresini et al. (2019) disapproved of their statements by mentioning that GCPSR is too conservative in detecting new species and also on several occasions failed to detect current speciation events as locus concordance occurs well after reproductive isolation. Contrarily, there are several past research evidences on sexual fertility capabilities among different *P. oryzae* lineages as well as on reproductive isolation between *P. oryzae* and *P. grisea* or *P. pennisetigena* (Kato et al., 2000; Couch & Kohn, 2002; Tosa et al., 2016). Moreover, the research findings of Gladieux et al. (2018) clearly revealed that gene flow and probabilistic chromosome painting resulted in the introgressions of chromosome segments from other host-specialised forms which led to the emergence of WB. Hence, there is continuing speciation in this fungus, and some of the lineages may end up fusing, hybridising or even disappearing. Similarly, the gene flow among wheat-infecting isolates and between wheat-infecting isolates and isolates infecting other host plants has already been reported by Inoue et al. (2017). Based on the above-mentioned points, it can be concluded that *Pygt* cannot be considered as a distinct species and hence it is better to define *Pygt* as *P. oryzae* pathotype *Triticum* as a valid name for the wheat blast fungus. It is important to note that *Triticum* strains have the potential to recombine in South Asia with sexually fertile *Oryza* strains in the Himalayan foothills (Saleh et al., 2014), which may result in the evolution of strains that may be aggressive on both rice and wheat. Fortunately, no *P. oryzae* lineages are currently adapted to barley or oats, but this could evolve in the near future if blast epidemics become common in these areas (Urashima et al., 2017). Thus, in reviewing all the published literature it seems that *M. oryzae* pathotype *Triticum* is the appropriate name for the wheat blast pathogen.

2.3 EVOLUTION AND EXPANSION OF WHEAT BLAST

Several laboratory and field experiments were conducted to define the mechanism and route of evolution of the wheat blast fungus. Preliminary experiments conducted by Urashima et al. (1993) revealed that fungal isolates causing the wheat blast disease have the potential to attack different types of native, invasive and cultivated grass hosts belonging to the Hordeae, Festuceae, Aveneae, Chlorideae, Agrosteae and Paniceae tribes. Later, Stukenbrock and McDonald (2008) provided evidence on the exchange of genetic materials among the *Pyricularia* populations infecting different plant hosts under field conditions due to physical contact between wheat and other grass hosts. Similar reports from Pereira et al. (2014) and Castroagudín et al. (2016) also concluded that cross-infection and inter-fertility between fungal isolates originating from different grass hosts growing around wheat fields play a crucial role in the emergence of wheat blast under Brazilian conditions. Recently, Inoue et al. (2017) provided genetic evidence in support of the hypothesis that the wheat blast fungus evolved through a host shift from a *Pyricularia* population infecting *Lolium*. Using a genetic model, they predicted that pwt3 strains were responsible for the development of the wheat blast epidemic in South America (Inoue et al., 2017). Through this model, the authors proposed that *Lolium*-derived isolates having the Ao avirulence allele at the PWT3 locus are able to cause infection in a susceptible wheat cultivar carrying the rwt3 allele as a susceptibility determinant and may be responsible for the wheat blast outbreak in the 1980s in South America as susceptible wheat cultivars carrying the *rwt3* gene are under widespread cultivation in the region. Further, they also mentioned that the selection pressure exerted by less popular Rwt3 wheat varieties triggers the rapid evolution of pathogen strains with non-functional PWT3 alleles. Comparative genome analysis revealed that MoT isolates in different wheat-growing regions of Bangladesh were clonal populations and were closely related to the highly aggressive MoT isolates of South America (Malaker et al., 2016; Farman et al., 2017). In an independent pathogenomics analysis performed by Islam et al. (2016), it was concluded that the wheat blast pathogen that appeared in Bangladesh most likely migrated from South America.

Historical records indicate that the MoT fungus was first noticed in 1985 in the Paraná state of Brazil (Igarashi et al., 1986) and the rapid spread of this fungus to Northern and Western Paraná, Northwestern São Paulo State and Southern Mato Grosso do Sul, Brazil was observed within a short time span of one year (Mottaleb et al., 2019a, 2019b) (Table 2.2). Its subsequent spread was observed in almost all the major wheat-growing areas of Brazil (Piccinini & Fernandez, 1989). Eleven years after its emergence in Brazil, the MoT fungus was recorded outside Brazil for the first time in Santa Cruz Department, Bolivia (Barea & Toledo, 1996). In 2002, the disease established itself in Itapúa and Alto Paraná Departments of Paraguay (Viedma, 2005), and five years later it occurred in the province of Formosa in Northeastern Argentina (Cabrera & Gutiérrez, 2007). The blast was observed in a small pocket in northeastern parts of Argentina (Alberione et al., 2008) and later observed on some wheat plants in Buenos Aires Province (Perelló et al., 2015). It was also observed on barley plants in the Corrientes Province of Argentina (Cabrera & Gutierrez, 2007). In 2016, WB was first noticed in mid-February 2016 in wheat fields

TABLE 2.2
Timeline Events of Expansion of Wheat Blast Pathogen at Global Scale

Year	Description	References
1985	First report of wheat blast in Paraná State, Brazil	Igarashi et al. (1986)
1986	Wheat blast spread and reported from São Paulo State and Mato Grosso do Sul State	Goulart et al. (1990)
1988	Wheat blast reported from Rio Grande do Sul State	
1990–1992	WB expanded to Minas Gerais and Goiás State in Brazil	Prabhu et al. (1992)
1993	WB reported from Brasília	Anjos et al. (1996)
1996	Observed in Eastern Bolivia in the lowlands of Santa Cruz Department	Barea and Toledo (1996)
2000–2003	WB disease expanded to Eastern Paraguay, i.e. Alto Paraná, Itapúa, Caaguazú, Caazapá Canindeyú and Guairá Departments	Viedma and Morel (2002); Viedma et al. (2010)
2007–2015	Northern Argentina (Chaco, Formosa and Corrientes Provinces) and Buenos Aires Province (experimental plots)	Perello et al. (2015)
2016	Southwestern Bangladesh (Barisal, Khulna and Rajshahi Divisions)	Malakar et al. (2016)

in the Chuadanga and Meherpur districts of Bangladesh, rapidly spreading to the adjacent eight districts of Bangladesh (Malaker et al., 2016; Mahmud, 2019; Sadat & Choi, 2017). The epidemic spread across an estimated 15,000 hectares, about 16% of the cultivated wheat area of Bangladesh, and yield losses reached 100% (Islam et al., 2016).

2.4 VIRULENCE SPECTRUM OF MoT FUNGUS

Several workers have studied the virulence spectrum of wheat-infecting MoT isolates in Brazil. Previous studies provided sufficient clues regarding the highly fertile nature of the MoT fungus (Urashima et al., 1993; Orbach et al., 1996; Tosa et al., 2016). These studies revealed that MoT isolates prevalent between 1980 and the 1990s were hermaphrodites and had the potential to cross with other pathotypes and produce perithecia and asci with 70%–95% viable ascospores. Hence, the role of sexual reproduction was prevalent in the early days of the wheat blast disease evolution and expansion.

The different isolates of the blast pathogen from various hosts showed differential infection potential and host specificity towards wheat cultivars (Peng et al., 2011). Urashima et al. (2004) also reported infection of *M. grisea* on triticale (*Secale* × *Triticum*) and barley (*Hordeum vulgare*). On cross-infectivity studies, it was observed that the blast pathogen from triticale and barley was able to infect triticale, barley, oat, wheat and rye, but not rice, sorghum, maize, common millet, sugarcane or *Brachiaria brizantha*. Zellerhoff et al. (2006) observed non-host-type resistance in barley against the *Magnaporthe* isolates derived from *Digitaria* (crabgrass), but not against the isolates from wheat, rice, maize or *Eleusine*. From the initial WB fungus in 1985, it only remained infective on wheat and black oats in Brazil. The blast disease was also detected in some other species of the Gramineae family, but white oats remained resistant until 2012 when it subsequently appeared in an epidemic on cultivar IAC 7 in São Paulo (Marangoni et al., 2013). It is worth mentioning that the first blast infection on perennial ryegrass (*Lolium perenne*) in the US in 2011 was a case of host jumping (Tosa et al., 2016).

Based on *in vitro* pathogenicity, reproductive isolation and genetic experimentation with different WB isolates, Urashima et al. (1993) established the existence of two different groups of isolates responsible for causing wheat blast in Brazil. According to the authors, one group has the potential to infect both rice and wheat, while the other group attacks only wheat. Further, they documented that wheat-derived isolates were also able to infect grassy weeds belonging to the Poaceae family. Additionally, they also provided evidence of fertile crosses among isolates based on cross-inoculation among isolates (wheat-derived isolates × isolates derived from different grasses such as *Eleusine coracana*, *Urochloa plantaginea* and *Setaria italic*) and the production of mature perithecia with viable ascospores (Urashima et al., 1993). However, crosses between wheat- and rice-derived isolates failed to produce fertile progenies while crosses between wheat- and weed-derived (*Setaria geniculata*, *Cenchrus echinatus* and *Echinocloa colonum*) isolates produced no perithecia (Urashima et al., 1993). Thus, all these research findings firmly established the role of a distinct *Pyricularia* population responsible for causing the WB disease in Brazil. Urashima and Kato (1994) evaluated 43 wheat line breeds, 7 *Triticum* spp. and 18 *Aegilops* lines from Brazil, Japan, the US and Bulgaria, against *M. grisea* under greenhouse inoculation conditions. Among these lines, only two *Aegilops* accessions were found resistant and all remaining were susceptible to wheat blast. To differentiate seedling and adult stage resistance, Cruz et al. (2010) challenge inoculated 70 different wheat genotypes at young leaf stage by employing 18 isolates of the MoT fungus. This study concluded that BRS229, BRS179, CNT8, BRS120 and BRS Buriti cultivars have a significant level of disease resistance towards the WB pathogen. However, at adult-stage inoculation, CNT8, NE 20156-B, PF 844001, PF 964009 and PF 804002 cultivars reflected less leaf and head area affected by the pathogen.

Urashima et al. (2005) also indicated the presence of two distinct populations (i.e. one sexually fertile and one infertile) responsible for causing the wheat blast epidemic in Brazil in 1998. The population from Paraná State was infertile. Molecular analysis based on DNA fingerprinting with MGR583 transposon sequences divided the Paraná isolates into two distinct clonal lineages. However, the majority of the isolates collected from the wheat field in Mato Grosso do Sul State were observed to be highly fertile mating-type hermaphrodites and showed strong isolate-to-isolate variation with the absence of a clear-cut demarcation for clonal lineages.

Maciel et al. (2014) comprehensively analysed 69 WB isolates using a detached wheat heads assay to predict cultivar resistance and their virulence spectrum. They noticed both partial and complete resistance among Brazilian wheat varieties at both seedling and adult stages of wheat. Partial resistance to the MoT fungus at the seedling stage was recorded in the varieties BRS 220, BR 18 (Terena) and BR 24, and at the heading stage in the case of the BRS 24 cultivar. Further, they also divided 69 MoT isolates into 14 seedling virulence group (SVGs) and provided evidences in favour of high frequencies of SVG A and the heading virulence group (HVG A′) as a result of a broad virulence range, and lower frequencies of SVGs F, G, M, P and R and HVGs S′, Q′ and T′ due to lack of virulence on some wheat cultivars. The uniform susceptibility of "Pleasant" barley indicates that barley may provide a key source of inoculum that could have led to the WB epidemic in Brazil. Similarly in Argentina, the aggressiveness of *P. oryzae* isolates on four commonly grown wheat cultivars under greenhouse conditions was studied and observations were made on lesions and the type of reaction at both the seedlings and spikes emergence stage (Martínez et al., 2019). At the seedling stage, Buck Meteoro displayed a susceptible reaction with seven isolates; however, the Baguette 11 cultivar displayed resistant reaction with all isolates. At the heading stage, the cultivars Klein Proteo and Baguette 11 showed 95% susceptibility reaction with isolates ArW22 and ArR1. However, minimum disease severity was recorded in cultivar ACA 303 and BolW8 when a combination of the isolates ArR3, ArR4 and BrW27 was used. The results clearly mentioned that a variation exists in the aggressiveness among *P. oryzae* isolates in Argentina.

An interesting observation related to the population genetics of the WB fungus was documented by Castroagudín et al. (2017) and Ceresini et al. (2018). During their studies, they identified a super race called virulence group A (VGA), which is responsible for causing blast on wheat and barley cultivars under artificial inoculation conditions. Further, they also demonstrated that the VGA phenotype develops as a result of major avirulence gene loss which is now in the segregation phase and present in different genetic backgrounds as a result of the recombination mechanism. This type of mechanism was not only observed at a significantly higher frequency in Brazilian wheat, but it was also noticed on several other weed grasses usually present in wheat-growing sessions. Major weed grasses include *Avena sativa*, *C. echinatus*, *Chloris distichophylla*, *Cynodon* spp., *Digitaria insularis*, *D. sanguinalis*, *Eleusine indica*, *Echinochloa crusgalli*, *Eragrostis plana*, *Panicum maximum*, *Rhynchelytrum repens*, *Sorghum sudanense* and *U. brizantha* (Ceresini et al., 2018). In addition to the VGA super race, other common and less virulent groups were also observed in several weedy grasses and wheat host attached by the WB fungal population. Therefore, this study provides additional clues for the effective movement or host jump of the WB fungus between wheat fields and nearby grasses. Overall, on the basis of Brazilian research experiences it can be concluded that

1. The wheat blast fungus is not a specialised wheat pathogen.
2. The hypothesis of grass-specific populations for the overall *P. oryzae* species complex is not true.
3. It is unclear whether the population derived from wheat and rice coexists and is responsible for causing the WB disease under natural conditions.
4. The rice blast pathogen cannot act as an appropriate model for understanding the ecology, epidemiology and population biology of the WB pathogen.

2.5 POPULATION STRUCTURE AND PATHOTYPE DIVERSITY

Several workers have investigated the biology and ecology of the wheat blast fungus (MoT) (Maciel et al., 2014; Urashima et al., 1999; Castroagudín et al., 2017). Various experimental evidences have suggested that the *M. oryzae* wheat-adapted population originated *de novo* from a non-rice host population. This is supported by the fact that both the populations *M. oryzae* pathotype *Triticum* (MoT) and *M. oryzae* pathotype *Oryza* (MoO) from Brazil have a clear-cut demarcation on the basis of DNA fingerprinting with repetitive DNA probes (MGR563 and MGR586) (Farman, 2002). In fact, the wheat-derived population showed more resemblance to isolates non-pathogenic to rice (Valent & Chumley, 1991; Hamer, 1991; Urashima et al., 1999; Farman, 2002). Similarly, Maciel et al. (2014) also observed that none of the 69 wheat-derived isolates of *M. oryzae* was able to infect rice. Thus, it was concluded that the Brazilian wheat-derived population was highly differentiated from the local rice-adapted population. In this study, they also employed microsatellite markers to explain the population structure of the MoT fungus in wheat fields of central-western, southeastern and southern parts of Brazil. Data analysis based on classical and coalescent-based historical migration revealed high levels of gene flow across a large spatial scale, as no clear demarcation in subdivision exists among the wheat-infecting populations. Further, they also observed that the clonal fraction of the MoT fungus was relatively high, irrespective of two mating-type idiomorphs (MAT1-1 and MAT1-2), which differ significantly in their frequencies. Based on these findings, they concluded that the WB fungus displayed a mixed reproductive system where initially reproduce by sexual means and further spread locally by the dispersal of an asexually produced clonal population. Overall, through this experimentation, it can be postulated that the high levels of gene flow among geographically distant populations are mainly due to (i) man-mediated long-distance transport of infested seed and (ii) airborne long-distance movement of pathogen propagules such as conidia. Urashima et al. (2007) have documented similar evidences on the dispersal of MoT conidia at least 1000 m from an infected field site.

Castroagudín and colleagues (2017) studied the population genetics of 500 isolates of the WB fungus obtained from wheat and grass species growing in or near infected fields across the wheat-growing regions of Brazil in 2012 and 2013. They observed high genetic and virulence diversity within the WB fungus with 198 multilocus microsatellite genotypes (MLMGs) and 25 virulence groups. Further, they mentioned that these populations displayed a mixed type of reproductive system, where sexual reproduction led to the production of novel genotypes, later selection depends on regional environmental conditions. The lateral spread and dispersal of well-adapted clones of this fungus took place as a result of asexual reproduction.

Gladieux et al. (2018) compared whole-genome sequences of 76 isolates of *M. oryzae*, which consisted of isolates from 12 host genera and 5 strains from related *Pyricularia* species for phylogenetic analysis. These were compared for polymorphism within lineages to levels of divergence between lineages or species to apprehend the relative evolutionary depth of the lineages within *M. oryzae*. In addition, whole-genome SNP data analysis using the pairwise distance method showed unequal distribution within and between the *M. oryzae* lineages. Analysis of gene flow showed that genetic exchanges have contributed to the makeup of multiple lineages within *M. oryzae*.

2.6 CONCLUSION AND FUTURE OUTLOOK

Population biology studies have enhanced current understanding of the adaptive evolution of the WB pathogen in an agro-ecosystem. Various studies conducted so far have highlighted that the WB fungus readily adapts to the agro-ecological environment due to higher rates of natural selection, genome-wide diversity and expansion. It has been documented that unexpectedly high levels of gene flow and probabilistic chromosome painting resulted in the introgressions of chromosome segments from other host-specialised forms which led to the emergence of wheat blast. There is continuous speciation in this fungus, and some lineages may end up fusing, hybridising or even disappearing. There is still a need to understand the mechanism that provides greater genome plasticity to adopt the agro-ecological environments by the WB pathogen. An equally important question that remains unanswered is to define further how the evolutionary capabilities of pathogen populations continue in spite of strong directional selection within homogeneous host environments. The identification of the *Avr* gene and the cognate resistance gene will help breeders to incorporate best R-gene combinations for effective MoT management. Hence, a deep understanding of the population dynamics and mechanisms for frequent mutation in MoT avirulence genes leading to new pathotypes will be an additional advantage for framing proper strategies for the management of the wheat blast disease.

In view of the urgent need to combat the WB disease in an effective and sustainable way, it is necessary to predict the emergence, adaptation and establishment of the wheat blast pathogen in agro-ecosystems by performing evolutionary genomics studies integrated with field experimental validations. A combination of evolutionary analysis of genome-wide patterns of genetic diversity in wheat and associated grassy weeds along with the pathogen by following integrated and targeted experiments linked with functional studies and the evolution of the WB fungus, will be imperative to channel the urgently required development of a novel, innovative and sustainable framework to slow down the emergence and spread of the WB fungus on a global scale.

REFERENCES

Alberione, E., C. Bainotti, I. Cettour and J. Salines. "Evaluación de enfermedades en trigos en siembra de verano en el NEA argentino-Campaña 2007/2008." In *7 Congreso Nacional de Trigo*. Santa Rosa, La Pampa, Argentina, 2008.

Anjos, J. R. N., D. B. Silva, M. J. D. Charchar and G. C. Rodrigues. "Occurrence of Blast Fungus (*Pyricularia grisea*) on Wheat and Rye in the Savanna Region of Central Brazil." *Pesquisa Agropecuária Tropical* 31 (1996): 79–82.

Araujo, L., J. M. Soares, M. C. C. Filippi and F. A. Rodrigues. "Cytological Aspects of Incompatible and Compatible Interactions Between Rice, Wheat and the Blast Pathogen *Pyricularia oryzae*." *Scientia Agricola* 73 (2016): 177–83.

Barea, G. and J. Toledo. "Identificacíon y zonificacíon de piricularia o bruzone (*Pyricularia oryzae*) en el cultivo del trigo en el Dpto. de Santa Cruz." In *Informe Técnico, Proyecto de Investigacíon Trigo*, 76–86. Cali, Columbia: CIAT, 1996.

Cabrera, M. G. and S. Gutiérrez. "Primer registro de *Pyricularia grisea* en cultivos de trigo del NE de Argentina." In *Jornada de Actualización en Enfermedades de Trigo*, edited by Lavallol, 60. Buenos Aires: IFSC Press, 2007.

Callaway, E. "Devastating Wheat Fungus Appears in Asia for First Time." *Nature* 532 (2016): 421–2.

Castroagudín, V. L., P. C. Ceresini, S. C. de Oliveira et al. "Resistance to Qoi Fungicides is Widespread in Brazilian Populations of the Wheat Blast Pathogen *Magnaporthe oryzae*." *Phytopathology* 105 (2015): 284–94.

Castroagudín, V. L., A. Danelli, S. I. Moreira et al. "The Wheat Blast Pathogen *Pyricularia graminis-tritici* has Complex Origins and a Disease Cycle Spanning Multiple Grass Hosts." *bioRxiv* (2017): DOI: 10.1101/203455.

Castroagudín, V. L., S. I. Moreira, D. A. S. Pereira et al. "Wheat Blast Disease Caused by *Pyricularia graminis-tritici* sp. Nov." *Persoonia* 37 (2016): 199–216. doi:10.3767/003158516X692149.

Ceresini, P. C., V. L. Castroagudín, F. A. Rodrigues et al. "Wheat Blast: Past, Present, and Future." *Annual Review of Phytopathology* 56 (2018): 427–56.

Ceresini, P. C., V. L. Castroagudín, F. A. Rodrigues et al. "Wheat Blast: From Its Origins in South America to its Emergence as a Global Threat." *Molecular Plant Pathology* 20, no. 2 (2019): 155–72.

Couch, B. C. and L. M. Kohn. "A Multilocus Gene Genealogy Concordant with Host Preference Indicates Segregation of a New Species, *Magnaporthe oryzae*, from *M. grisea*." *Mycologia* 94 (2002): 683–93.

Cruz, C. D., J. Kiyuna, W. W. Bockus, T. C. Todd, J. P. Stack and B. Valent. "*Magnaporthe oryzae* Conodia on Basal Wheat Leaves as a Potential Source of Wheat Blast Inoculum." *Plant Pathology* 64 (2015): 1491–8.

Cruz, M. F. A., A. M. Prested, J. L. Maciel and P. L. Scheeren. "Partial Resistance to Blast on Common and Synthetic Wheat Genotypes in Seedling and in Adult Plant Growth Stages." *Tropical Plant Pathology* 35 (2010): 24–31.

Farman, M., G. L. Peterson, L. Chen et al. "The *Lolium* Pathotype of *Magnaporthe oryzae* Recovered from a Single Blasted Wheat Plant in the United States." *Plant Disease* 101 (2017): 684–92.

Farman, M. L. "*Pyricularia grisea* Isolates Causing Gray Leaf Spot on Perennial Ryegrass (*Lolium perenne*) in the United States: Relationship to *P. grisea* Isolates from Other Host Plants." *Phytopathology* 92 (2002): 245–54.

Filha, M. S. X., F. A. Rodrigues, G. P. Domiciano, H. V. Oliveira, P. R. Silveira and W. R. Moreira. "Wheat Resistance to Leaf Blast Mediated by Silicon." *Australasian Plant Pathology* 40 (2011): 28–38.

Gladieux, P., B. Condon, S. Ravel et al. "Gene Flow between Divergent Cereal- and Grass-Specific Lineages of the Rice Blast Fungus *Magnaporthe oryzae*." *mBio* 9, no. 1 (2018): e01219–17.

Goulart, A. C. P., F. A. Paiva and A. N. Mesquita. "Occurrence of Wheat Blast (*Pyricularia oryzae*) in the State of Mato Grosso do Sul." *Fitopatologia Brasileira* 15 (1990): 112–14.

Hamer, J. E. "Molecular Probes for Rice Blast Disease." *Science* 252 (1991): 632–3.

Igarashi, S., C. M. Utiamada, L. C. Igarashi, A. H. Kazuma and R. S. Lopes. "*Pyricularia emtrigo*. 1. Ocorrência de *Pyricularia* sp. no estado do Parana." *Fitopatologia Brasileira* 11 (1986): 351–2.

Inoue, Y., T. T. P. Vy, K. Yoshida et al. "Evolution of the Wheat Blast Fungus Through Functional Losses in a Host Specificity Determinant." *Science* 357 (2017): 80–3.

Islam, M. T., C. Daniel, P. Gladieux et al. "Emergence of Wheat Blast in Bangladesh was Caused by a South American Lineage of *Magnaporthe oryzae*." *BMC Biology* 14 (2016): 84–94.

Islam, M. T., K.-H. Kim and J. Choi. "Wheat Blast in Bangladesh: The Current Situation and Future Impacts." *The Plant Pathology Journal* 35 (2019): 1–10.

Kashyap, P. L., S. Kumar, R. Tripathi et al. "Phylogeography and Population Structure Analysis Reveals Diversity by Gene Flow and Mutation in *Ustilago segetum* (Pers.) Roussel *tritici* Causing Loose Smut of Wheat." *Frontiers in Microbiology* 10 (2019): 1072.

Kato, H., M. Yamamoto, T. Yamaguchi-Ozaki et al. "Pathogenicity, Mating Ability and DNA Restriction Fragment Length Polymorphisms of *Pyricularia* Populations Isolated from Gramineae, Bambusideae and Zingiberaceae Plants." *Journal of General Plant Pathology* 66 (2000): 30–47.

Kohli, M. M., Y. R. Mehta, E. Guzman, L. De Viedma and L. E. Cubilla. "*Pyricularia* Blast—A Threat to Wheat Cultivation." *Czech Journal of Genetics and Plant Breeding* 47 (2011): S130–4.

Luikart, G., P. R. England, D. Tallmon, S. Jordan and P. Taberlet. "The Power and Promise of Population Genomics: From Genotyping to Genome Typing." *Nature Reviews Genetics* 4 (2003): 981–94.

Maciel, J. L. N., P. C. Ceresini, V. L. Castroagudın, G. H. J. Kema and B. A. McDonald. "Population Structure and Pathotype Diversity of the Wheat Blast Pathogen *Magnaporthe oryzae* 25 Years After its Emergence in Brazil." *Phytopathology* 104 (2014): 95–107.

Mahmud, H. "Wheat Blast in Bangladesh Threatening South Asia Wheat Production." *Acta Scientific Microbiology* 2, no. 1 (2019): 8–9.

Malaker, P. K., N. C. D. Barma, T. P. Tiwari et al. "First Report of Wheat Blast Caused by *Magnaporthe oryzae* Pathotype *Triticum* in Bangladesh." *Plant Disease* 100 (2016): 2330.

Marangoni, M. S., M. P. Nunes, N. Fonseca Junior and Y. R. Mehta. "*Pyricularia* Blast on White Oats: A New Threat to Wheat Cultivation." *Tropical Plant Pathology* 38 (2013): 198–202.

Martínez, S. I., A. Sanabriaa, M. C. Fleitasa, V. F. Consolo and A. Perelló. "Wheat Blast: Aggressiveness of Isolates of *Pyricularia oryzae* and Effect on Grain Quality." *Journal of King Saud University – Science* 31 (2019): 150–7.

Mottaleb, K. A., P. K. Singh, K. Sonder et al. "Threat of Wheat Blast to South Asia's Food Security: An Ex-Ante Analysis." *PLoS ONE* 13, no. 5 (2018): e0197555. doi:10.1371/journal.pone.0197555.

Mottaleb, K. A., V. Govindan, P. K. Singh et al. "Economic Benefits of Blast-Resistant Biofortified Wheat in Bangladesh: The Case of BARI Gom 33." *Crop Protection* (2019b). doi:10.1016/j.cropro.2019.05.013.

Mottaleb, K. A., P. K. Singh, X. He, A. Hossain, G. Kruseman and O. Erenstein. "Alternative Use of Wheat Land to Implement a Potential Wheat Holiday as Wheat Blast Control: In Search of Feasible Crops in Bangladesh." *Land Use Policy* 82 (2019a): 1–12.

Oliveira, S. C., V. L. Castroagudın, J. L. N. Maciel, D. A. dos Santos Pereira and P. C. Ceresini. "Cross Resistance to QoI Fungicides Azoxystrobin and Pyraclostrobin in the Wheat Blast Pathogen *Pyricularia oryzae* in Brazil." *Summa Phytopathologica* 41 (2015): 298–304.

Orbach, J., F. Chumley and B. Valent. "Electrophoretic Karyotypes of *Magnaporthe grisea* Pathogens of Diverse Grasses." *Molecular Plant-Microbe Interactions* 9 (1996): 261–71. doi:10.1094/MPMI-9-0261.

Ou, S. H. *Fungus Diseases – Foliage Diseases. Rice Diseases.* 2nd ed. 109–201. Kew: CAB, 1985.

Pagani, A. P. S., A. C. Dianese and A. C. Café-Filho. "Management of Wheat Blast with Synthetic Fungicides, Partial Resistance and Silicate and Phosphite Minerals." *Phytoparasitica* 42 (2014): 609–17.

Peng, J. L., Y. L. Zhou and Z. H. He. "Global Warning Against the Spread of Wheat Blast." *Journal of Triticeae Crops* 31 (2011): 989–93.

Pereira, J. F., L. Consoli, E. A. S. Bombonatto, A. L. V. Bonato and J. L. N. Maciel. "Development of Genomic SSR Markers and Molecular Characterization of *Magnaporthe oryzae* Isolates from Wheat in Brazil." *Biochemical Genetics* 52 (2014): 52–70.

Perelló, A., I. Martinez and M. Molina. "First Report of Virulence and Effects of *Magnaporthe oryzae* Isolates Causing Wheat Blast in Argentina." *Plant Disease* 99, no. 8 (2015): 1177.

Piccinini, E. C. and J. M. C. Fernandez. "Ocorrência da brusone (*Pyricularia oryzae*) em lavouras comerciais de trigo (*Triticum aestivum*) no estado do Rio Grande do Sul." *Fitopatologia Brasileira* 14, no. 2 (1989): 126. (Abstract: 1989.)

Prabhu, A. S., M. C. Filippi and N. Castro. "Pathogenic Variation Among Isolate of *Pyricularia oryzae* Infecting Rice, Wheat and Grasses in Brazil." *Tropical Pest Management* 38 (1992): 367–71.

Rocha, J., A. Pimentel, G. Ribeiro and M. De Souza. "Eficiência de fungicidas no controle da brusone em trigo." *Summa Phytopathologica* 40 (2014): 347–52.

Rocha, J. R. A. S. C., I. G. de Paula, H. B. Gloria, A. J. B. Pimentel, G. Ribeiro and M. A. de Souza. "Screening Wheat Genotypes for Resistance to Wheat Blast Disease in the Vegetative and Reproductive Stages." *Euphytica* 215 (2019): 59.

Sadat, M. D. and J. Choi. "Wheat Blast: A New Fungal Inhabitant to Bangladesh Threatening World Wheat Production." *Plant Pathology Journal* 33 (2017): 103–8.

Saleh, D., J. Milazzo, H. Adreit, E. Fournier and D. Tharreau. "South-East Asia is the Center of Origin, Diversity and Dispersion of the Rice Blast Fungus, *Magnaporthe oryzae*." *Molecular Plant Pathology* 201, no. 4 (2014): 1440–56.

Silva, W. L., M. F. A. Cruz, A. A. Fortunato and F. A. Rodrigues. "Histochemical Aspects of Wheat Resistance to Leaf Blast Mediated by Silicon." *Scientia Agricola* 72 (2015): 322–7.

Singh, D. P. "Wheat Blast Caused by *Magnaporthe oryzae* pathotype *Triticum*: Present Status, Variability, and Strategies for Management." In *Management of Wheat and Barley Diseases*, edited by D. P. Singh, 635–43. Waretown, NJ: Apple Acad, 2017.

Stukenbrock, E. H. and B. A. McDonald. "The Origins of Plant Pathogens in Agro-Ecosystems." *Annual Review of Phytopathology* 46 (2008): 75–100.

Tosa, Y., Y. Inoue, T. P. V. Trinh and I. Chuma. "Genetic and Molecular Analysis of the Incompatibility Between *Lolium* Isolates of *Pyricularia oryzae* and Wheat." *Physiological and Molecular Plant Pathology* 95 (2016): 84–6.

Urashima, A. S. and H. Kato. "Varietal Resistance and Chemical Control of Wheat Blast Fungus." *Summa Phytopathologica* 20 (1994): 107–12.

Urashima, A. S., R. Galbieri and A. Stabili. "DNA Fingerprinting and Sexual Characterization Revealed Two Distinct Populations of *Magnaporthe grisea* in Wheat Blast from Brazil." *Czech Journal of Genetics and Plant Breeding* 41 (2005): 238–45.

Urashima, A. S., S. Igarashi and H. Kato. "Host Range, Mating Type and Fertility of *Pyricularia grisea* from Wheat in Brazil." *Plant Disease* 77 (1993): 1211–6.

Urashima, A. S., S. F. Leite and R. Galbieri. "Efficiency of Aerial Dissemination of *Pyricularia grisea*." *Summa Phytopathologica* 33, no. 3 (2007): 275–9.

Urashima, A. S., A. F. Alves, F. N. Silva, D. Oliveira and R. Gazaffi. "Host Range, Mating Type and Population Structure of *Magnaporthe* sp. of a Single Barley Field in São Paulo State, Brazil." *Journal of Phytopathology* 165 (2017): 414–24.

Urashima, A. S., C. Grosso, A. Stabili et al. Effect of *Magnaporthe grisea* on Seed Germination, Yield and Quality of Wheat. In Wang, G. L. and Valent, B. (eds.) *Advances in Genetic, Genomics and Control of Rice Blast Disease*, 267–277. Netherlands: Springer, 2009.

Urashima, A. S., N. A. Lavorent, A. C. P. Goulart and Y. R. Mehta. "Resistance Spectra of Wheat Cultivars and Virulence Diversity of *Magnaporthe grisea* Isolates in Brazil." *Fitopatologia Brasileira* 29 (2004): 511–8.

Urashima, A. S., Y. Hashimoto, Y. Don, M. Kusaba, Y. Tosa, H. Nakayashiki and S. Mayama. "Molecular Analysis of the Wheat Blast Population in Brazil with a Homolog of Retrotransposon MGR583." *Annals of the Phytopathological Society of Japan* 65 (1999): 429–36.

Valent, B. and F. G. Chumley. "Molecular Genetic Analysis of the Rice Blast Fungus, *Magnaporthe grisea*." *Annual Review of Phytopathology* 29 (1991): 443–67.

Valent, B., C. D. Cruz, M. Farman et al. "Strategies for Managing Blast Disease of Wheat." *Achieving Durable Resistance to Wheat Diseases and Pests*, 5, 2–3 November 2016, Bloomington, MN.

Valent, B., M. Farman, Y. Tosa et al. "*Pyricularia graminis-tritici* is Not the Correct Species Name for the Wheat Blast Fungus: Response to Ceresini et al. (MPP 20:2)." *Molecular Plant Pathology* 20, no. 2 (2019): 173–9.

Viedma, L. Q. "Wheat Blast Occurrence in Paraguay (abstract)." *Phytopathology* 95, no. 6s (2005): S152.

Viedma, L. Q. and W. Morel. *Añublo o Piricularia del Trigo (Wheat Blast).* Díptico. Capitán Miranda, Itapúa, Parag: MAG/DIA/CRIA, 2002.

Viedma, L. Q., M. M. Kohli, L. Cubilla and G. Cabrera. "Manejo integrado de mancha amarilla y la Piricularia en el cultivo de trigo en Paraguay." In *Tercer Semin. Nacional de Trigo: "Del Grano al Pan"*, edited by M. Kohli, 31–42. LE Cubilla: G Cabrera. Asuncíon, Parag.: Capeco/Inbio 159, 2010.

Wang, S., S. Asuke, T. T. P. Vy et al. "A New Resistance Gene in Combination with Rmg8 Confers Strong Resistance Against *Triticum* Isolates of *Pyricularia oryzae* in a Common Wheat Landrace." *Phytopathology* 108 (2018): 1299–306. doi:10.1094/PHYTO-12-17-0400-R.

Yoshida, K., D. G. Saunders, C. Mitsuoka et al. "Host Specialization of the Blast Fungus *Magnaporthe oryzae* is Associated with Dynamic Gain and Loss of Genes Linked to Transposable Elements." *BMC Genomics* 18 (2016): 370.

Zellerhoff, N., B. Jarosch, J. Z. Groenewald, P. W. Crous and U. Schaffrath. "Nonhost Resistance of Barley is Successfully Manifested Against *Magnaporthe grisea* and a Closely Related Pennisetum-Infecting Lineage but is Overcome by *Magnaporthe oryzae*." *Molecular Plant-Microbe Interactions* 19 (2006): 1014–22.

Zhang, N., J. Luo, A. Y. Rossman et al. "Generic Names in Magnaporthales." *IMA Fungus* 7 (2016): 155–9.

Zhou, B., S. Qu, G. Liu, M. Dolan, H. Sakai, G. Lu, M. Bellizzi, and G. L. Wang. "The Eight Amino-Acid Differences within Three Leucine-Rich Repeats between Pi2 and Piz-t Resistance Proteins Determine the Resistance Specificity to *Magnaporthe grisea*." *Molecular Plant-Microbe Interactions* 19 (2006): 1216–28.

3 Identification and Diagnosis of Wheat Blast

Prem Lal Kashyap, Sudheer Kumar, Krishna Gopal, Sunita Mahapatra, Satyajit Hembram and Gyanendra Pratap Singh

CONTENTS

3.1 INTRODUCTION

Wheat blast (WB) is caused by *Magnaporthe oryzae* (synonymous with *Pyricularia oryzae*) pathotype *Triticum* (MoT) (Gladieux et al., 2018; Zhang et al., 2016; Couch & Kohn, 2002). It is reported as one of the principle and most terrifying diseases of wheat under congenial conditions and results in total crop failure (Sadat & Choi, 2017; Callaway, 2016; Saharan et al., 2016; Maciel et al., 2014; Cardoso et al., 2008; Urashima et al., 2009; Cruz & Valent, 2017). Up to 2016, the MoT fungus was confined to South America before its widespread outbreak in Bangladesh (Islam et al., 2016; Malaker et al., 2016). Although *M. oryzae* pathotypes exhibit a high degree of host specificity, they are reported to infect black oat, rye and barley, as well as non-cultivated graminaceous hosts (Mehta & Baier, 1998; Mehta et al., 2006; Marangoni et al., 2013; Anjos & Charchar, 2000; Kohli et al., 2011). Wheat blast is characterised by bleaching of the head, which closely resembles *Fusarium* head blight (FHB) symptoms and also *Bipolaris* infection, thus the identification and monitoring of wheat blast at the field level become complicated (Singh, 2017a; Valent et al., 2013). Wheat blast can cause up to 100% yield loss under favourable conditions

(Martínez et al., 2018; Cruz et al., 2012; Kohli et al., 2011). Higher yield losses occur when favourable weather conditions coincide with the vulnerable growth stages of the host during infection (Urashima et al., 2009). Several control measures, including prophylactic methods and fungicide applications are currently used to control WB. The application of chemical fungicide at the ear heading stage is a cost-effective and easy way to reduce the inoculum size of the pathogens. Seed treatments with fungicide will be helpful in eliminating the seed-borne infection (Bockus et al., 2015). Because wheat blast symptoms mainly appear on the spikes, fungicides combining triazoles with strobilurins have been suggested to control the disease (Kohli et al., 2011). However, the development of resistance against fungicides is another challenging issue for the chemical control of the disease. The resistance of the wheat blast pathogen to strobilurin fungicides (quinoneoutside inhibitors) increased from 36% of the Brazilian population in 2005 to 90% in 2012 (Castroagudín et al., 2015). Alternatively, a combination of silicon or phosphite with fungicide treatment reduced the disease on the spikes (Pagani et al., 2014). The discovery of a more effective fungicide with a novel mode of action is necessary. Under high disease pressure, wheat varieties are also unable to provide complete resistance towards WB (Ceresini et al., 2018). As MoT is efficient at jumping from one grass host to another and can spread through various ways including trade and wind, rapid, sensitive and reliable methods to understand the inoculum load of the pathogen present in the seed and soil, on weeds, collateral hosts, etc., become crucial for timely and effective disease management of the WB disease. Commonly, a visual assessment of the symptoms, lesion diameter measurement, spore and other pathogen propagules present at onset of infection are being used for the diagnosis of diseases. However, these tests are tedious, time-consuming and often cannot discriminate between closely related plant pathogens that produce similar symptom. Besides, the tests are not well suited to assess fungal development in the early phases of the infection, before macroscopic symptoms are visible or before sporulation. Therefore, this chapter presents and discusses an updated and detailed review of the detection and diagnosis of the MoT fungus.

3.2 PHENOTYPIC IDENTIFICATION OF MoT

Wheat blast is considered one of the most terrifying diseases of wheat (Figure 3.1) due to its sporadic nature. The symptoms of MoT infection occur on all aerial wheat tissues, such as the leaves, stems, spikes and rachises, and, in the later stages of the disease, the infected spikes are bleached (Igarashi, 1990) (Table 3.1). A detailed description of the different symptoms produced by the fungus on wheat plants is provided in the following subheadings.

3.2.1 Symptoms on Leaf and Stem

The MoT fungus develops symptoms on the leaves, stem and the junction of the leaf blade and leaf sheath of wheat (Monsur et al., 2016). Goulart and Paiva (1990) observed that with the emergence of seedlings, the MoT fungus colonises new tissues including the coleoptiles, stem and primary leaves. Under favourable conditions,

FIGURE 3.1 Blast-infected farmer's field in the Meherpur district of Bangladesh showing severe blast infection with silvery white spikes and green canopy. (Courtesy: Pawan Singh, CIMMYT, Mexico.)

TABLE 3.1
Symptomatic Difference between Blast and Head Scab Infection in Wheat

Criteria of differentiation	Blast	Head scab
Spike bleaching	More creamy but with grey MoT spore masses at the base of the bleached section	Traces of orange to pink spore masses
Infection point or area within single head	Single	Multiple
Infection movement from point of infection	Upwards	Both upwards and downwards

MoT has the potential to produce numerous conidia on the basal leaves, and the production of conidia on leaves coincides with wheat spike emergence (Cruz et al., 2015). Thus, MoT not only causes lesions on leaves at the seedling stage (three-leaf stage), but also sporulates on the wheat basal leaves of adult plants in field environments. On leaves, initial macroscopic lesions appear as water soaked to grey-green. The centre of these blast lesions turns grey and white to tan during and after sporulation, respectively. Generally, mature lesions show a dark brown to reddish-brown margin with yellow chlorotic halos (Figure 3.2). Individual lesions appear as an eye-spot with an elliptical shape. Sometimes, lesions coalesce in moderate to severely infected seedlings, resulting in the complete destruction of the plant (Rios et al., 2013). However, under field conditions, the symptoms on leaves are difficult

FIGURE 3.2 Wheat blast symptoms on (A) young leaf and (B) mature wheat leaf. (Courtesy: Pawan Kumar Singh, CIMMYT, Mexico.)

to identify because of the mixed infection of spot blotch. In comparison to rice blast infection, when larger lesions occur on younger rice leaves, blast lesions in wheat occur more frequently on older leaves, including senescent leaves at the base of the plant (Cruz et al., 2015). In rare cases, lesions are also seen on the leaf collar, culm, culm nodes and stem (Figure 3.3). Severe wheat blast outbreaks without blast symptoms on leaves are related to tissue-specific symptoms caused by microconidia transported to the wheat heads (Zhang et al., 2014).

3.2.2 Symptoms on Spike and Wheat Head

The spike blight symptoms of wheat blast are often confused with *Fusarium* head scab (Valent et al., 2013) and spike blight due to spot blotch in the field under a humid and warm climate. When the rachis is infected with *Fusarium* head blight, spikelets above and below the point of infection become bleached (Wise et al., 2015), and the infected spikelets contain pink to orange masses of spores (Farman et al., 2017; Valent et al., 2016; Wise et al., 2015). In the case of spike blight due to *Bipolaris sorokiniana*, dark brown discolouration develops on the spike and such spikes may possess healthy spikelets at both ends from the point of infection (Singh, 2017a). However, in MoT infections, bleached yellow or white heads

FIGURE 3.3 Wheat blast symptoms on stem. (Courtesy: Pawan Kumar Singh, CIMMYT, Mexico.)

(Figure 3.4) appear (Farman et al., 2017). Further, the MoT fungus attacks the rachis, while the head blight pathogen attacks individual spikelets (Duveiller et al., 2012). Wheat heads turn white or tan from the point of infection upwards and produce greyish mould at the point of attachment of individual spikelets (Figure 3.4). Contrarily, the diseased head lacks the typical pinkish tinge which is characteristic of sporulation of the *Fusarium* head blight fungus. Generally, head bleaching occurs in the absence of leaf symptoms (Urashima et al., 2009; Kohli et al., 2011; Cruz et al., 2012) and is the most prominent and easily identifiable symptom of wheat blast under field conditions (Figure 3.1). Most predominately, the fungus produces a dark and brilliant lesion composed of a grey and sporulating mycelial mass at the point of infection in rachis (Kashyap et al., 2018; Lau et al., 2011). Symptoms on heads can vary from elliptic lesions and bleached centres to spike bleaching, sterility and empty grains, depending on the time of infection (Saharan et al., 2016). Significant yield losses have been noticed when spike infections occur during the flowering or early stage of grain formation (Goulart et al., 2007). This is due to the fact that infection in the rachis or peduncle can obstruct the translocation of photosynthates and destroy the upper parts of the spike (Figure 3.2). As a result, partial or total spike sterility can occur depending on the point of infection, the timing of infection and the resistance level of the cultivar. In severe cases, the

FIGURE 3.4 Wheat blast (MoT) symptoms on (A) awns, (B) spikelet, (C) spike and (D) rachis. (Courtesy: Pawan Kumar Singh, CIMMYT, Mexico.)

entire head is damaged. If head infection occurs early, then grain production can drastically decline.

3.2.3 Symptoms on Wheat Awns

The infected glumes show elliptical lesions with reddish-brown to dark grey margins while awns show brown to whitish discoloration (Figure 3.4). Lesions have grey centres during sporulation and white to tan centres after spore release (Igarashi, 1990; Igarashi et al., 1986).

3.2.4 Symptoms on Wheat Grains and Seeds

MoT is a seed-borne fungus and can be present in asymptomatic seeds. Thus, infected seeds provide the initial inoculum to new areas where infected seeds are grown (Toledo & Escobar, 2002; Urashima & Kato, 1994). It has been noticed that MoT can spread from spike to seed and from infected seeds to seedlings. MoT infections that occur late in the season under favourable conditions may increase the chance of seed transmission of the fungus, resulting in infected seeds (Igarashi, 1990). Goulart and Paiva (1990) estimated that the rate of MoT dissemination from a non-treated seed lot with 21% incidence could potentially produce 400,000 primary inoculum units per hectare. The grains of MoT-infected spikes from highly susceptible cultivars are usually deformed, shrivelled and small in size with a low test weight (Goulart et al., 2007). Small, shrivelled seeds can be mixed with normal-appearing seeds on symptomatic spikes while threshing (Urashima et al., 2009).

3.3 MICROSCOPIC DETECTION OF MoT

The lesions that develop on spikes and leaves possess three-celled pyriform conidia on aerial conidiophores that can be seen under a microscope. The MoT fungus produces three different kinds of spores on a wheat host. These include (i) asexual pyriform conidia; (ii) asexual microconidia produced from phialides; and (iii) perithecia and ascospores (Cruz et al., 2015). The culture of the MoT fungus is greyish in appearance on a V8 juice agar media. A moist blotter assay indicated that the infected plant parts produce conidia when incubated on a three-layered moist blotter for 12 h light/darkness cycle at ambient conditions for five days (Malakar et al., 2016). The MoT fungus produces hyaline to pale grey–coloured pyriform-shaped asexual conidia (Figure 3.5). The conidiophores are single or clustered, simple, infrequently branched and exhibit sympodial growth. At the tip of the conidiophore, conidia form at points that ascend sympodially and in succession, narrowed towards the tip, pyriform to obclavate, rounded at the base, 2-septate, rarely 1- or 3-septate, hyaline to pale olive (Singh, 2017b).

3.4 MOLECULAR DIAGNOSIS OF MoT

The accurate detection of the presence of a pathogen even in very low quantities before disease symptoms is essential in devising appropriate and timely disease management tactics (Sharma et al., 2017; Kashyap et al., 2011). In such situations, recent developments in DNA barcoding technology are quite helpful in detecting and identifying the presence of specific pathogens, providing a practical, specific and sensitive identification tool that can be used for diagnostics, epidemiological and ecological studies as well as for monitoring fungal plant pathogens (Kashyap et al., 2017b). Further, nucleic acid–based pathogen detection technologies can overcome the limitations of traditional and morphological pathogen diagnosis assays with the use of DNA amplification for species-specific and strain-specific detection of microbes (Chakdar et al., 2019; Sharma et al., 2017). Basically, polymerase chain reaction (PCR) offers a simple and nifty procedure for the exponential amplification of nucleic acids (DNA sequences) by *in vitro* DNA synthesis. There are three steps in PCR amplification: (i) melting of the target DNA segment, (ii) annealing of primers (forward and reverse primers) to the denatured DNA strands and (iii) primer extension by DNA polymerase (Kashyap et al., 2016). Newly synthesised DNA strands serve as targets for subsequent PCR cycles in three steps and these cycles should be repeated up to 40 times. Thus, PCR provides several advantages over traditional methods of pathogen diagnosis. For instance, the pathogens do not need to be cultured prior to their diagnosis by PCR. Further, theoretically, the technique has the potential to detect a single molecule of the target without using radioactive probes.

FIGURE 3.5 Spore morphology of MoT fungus under (A) compounds microscope and (B) scanning electron microscope. (Courtesy: Pawan Kumar Singh, CIMMYT, Mexico.)

The technique is rapid and versatile, and through multiplexing can detect two to four targets in a single run using multiple primer sets. Similar to serology, both narrow and broad selectivity are possible but at a lower cost in comparison to serological reagent development and assays. The synthesis of hundreds of different PCR primers generates costs comparable to those of developing only a few monoclonal antibodies. The following sections provide detailed information on the different types of molecular diagnostic assays developed so far for MoT detection and diagnosis.

3.4.1 Nucleic Acid Barcodes

PCR-based diagnostic assays based on nucleic acid barcodes are being used and established in methods for the detection of plant pathogens in several diagnostic laboratories and regulatory agencies. Such assays enable the detection of the presence of a pathogen even in the absence of any signs or symptoms of the pathogen especially during the early stages of infection or latent infection. Generally, regions from conserved genes such as the ribosomal DNA (rDNA) internal transcribed spacer (ITS) regions 1 and 2, translation elongation factor 1-alpha and β-tubulin are utilised in the development of specific primers (Kashyap et al., 2017b; Vincelli & Tisserat, 2008). PCR assays based on the ITS regions are favoured because ITS regions 1 and 2 have a conserved region as well as considerable variation among individual fungal species and are thus able to differentiate between closely related species (White et al., 1990; Kashyap et al., 2017b). Additionally, rDNA has been found in a number of copies in the genome of fungal pathogens; therefore, ITS-based PCR markers are more sensitive in comparison to single-copy genes–based markers (Vincelli & Tisserat, 2008). With the advent of new and powerful genome sequencing technologies, it has now become cost-effective to sequence entire genomes of closely related plant pathogens to rapidly obtain targets for diagnostic assay development (Lang et al., 2010, 2014). The advantages of this approach are that prior knowledge regarding target sites is not required and that it allows for the simultaneous development of markers that can then be empirically tested for specificity.

M. oryzae–specific transposable element sequences such as Pot2, MGR583 and MoTeR have been useful in the identification of *M. oryzae*, but they do not differentiate MoT isolates from the isolates of other host-specific pathotypes (Farman, 2002). As a result, Pieck et al. (2017) employed a genomics-based approach to develop molecular markers specific to MoT fungus (Table 3.2). They identified regions suitable for selecting specific primers for the development of a PCR-based assay utilising whole-genome draft sequences of *M. oryzae* strains and a bioinformatic pipeline. The assays developed were able to differentiate between the MoT strains present in South America and other *M. oryzae* pathotypes. In stepwise order, they sequenced and assembled the genomes of multiple MoT strains followed by whole-genome comparisons to identify sequences sufficiently unique to the MoT pathotype for DNA barcode development and validation. Based on their bioinformatic analysis, they developed the MoT3 marker matching the WB12 sequence corresponding to a fragment of the *M. oryzae* MGG_02337 gene annotated as a short-chain dehydrogenase for the development of a PCR-based diagnostic assay. Further, they tested the specificity of the developed marker against a panel of DNA extracted from 284 *M. oryzae* isolates collected from a diverse array of host species including the MoT fungus. The selected primer set amplified a 361-bp product in a conventional PCR, and it was able to amplify from as little as 0.1 ng of purified DNA. The specificity of the MoT3-based assay was also evaluated using *Fusarium* spp. DNA, where no amplicon generation was detected. Contrarily, a study conducted by Gupta et al. (2019) revealed that the MoT3 assay could not reliably distinguish between wheat and rice blast isolates from Bangladesh based on DNA amplification experiments. Recently, in response to this study, a note was submitted in which Yasuhara-Bell et al. (2018) pointed out that the assay described in the original paper by Pieck et al. (2017) is able to reliably distinguish between wheat and rice isolates from Bangladesh and worldwide. In such circumstances, it is our opinion that further research efforts are still required to develop a convenient diagnostic tool for the detection of wheat blast from a suspected seed lot, a asymptomatic plant and alternative hosts.

TABLE 3.2
List of Specific DNA Regions and Primer Sequences Used in the Nucleic Acid–driven Detection of MoT Fungus on Wheat

Sequence ID/ amplicon size	Primer sequence	Assay type	References
MoT3_F1	AGGATGTATGCCCTGACTGG	Conventional PCR	Pieck et al. (2017)
MoT3_R1	CTCGGCGATGCAAAAGTGAA	Conventional PCR	Pieck et al. (2017)
MoT3_FAM2	5′-[6FAM]-TTCTAACGGTTTGCAATTGCAGAAACAAC-[BHQ1]-3′	Real-time PCR	Pieck et al. (2017)
PoT2-F3	AATTACGCTTTTTTGCCGA	LAMP	Yasuhara-Bell et al. (2018)
PoT2-B3	GGTTTAATAGCGTTAATAACCGG	LAMP	Yasuhara et al. (2018)
PoT2-FIP	GAGCCAAAAAACGAGTTATCCAACGACGAATTCTTCAAGCCGC	LAMP	Yasuhara et al. (2018)
PoT2-BIP	AATCCTTAAAACCCAAAGGCCCACCAAGATTTAATTACCTCCGT	LAMP	Yasuhara et al. (2018)
PoT2-LoopF	CCAAGGCCTTTTGTCTCTCCG	LAMP	Yasuhara et al. (2018)
PoT2-LoopB	CGTCGAATAGATAACGCCCGG	LAMP	Yasuhara et al. (2018)
PoT2-Probe/56-ROXN/	ACGCTGAGGACCCGGATGCGAATGCGGATGCGGATGCCGACGTCGAATAGATAACGCCCGG	LAMP	Yasuhara et al. (2018)
MoT3-F3	CCAAGGATGTATGCCCTGAC	LAMP	Yasuhara et al. (2018)
MoT3-B3	GTTGGGGGCTTCGTATGC	LAMP	Yasuhara et al. (2018)
MoT3-FIP	CAAACCGTTAGAAAGCTGTACTTCAGTTGTTACAACATCACGAATCG	LAMP	Yasuhara et al. (2018)
MoT3-BIP	TTGCACAAAACAACCCAGTTCAGAGACGTTGTTTATGAATTCCG	LAMP	Yasuhara et al. (2018)
MoT3-LoopB	CTTTTGCATCGCCGAGCC	LAMP	Yasuhara et al. (2018)
MoT3-Probe*/5Cy5	ACGCTGAGGACCCGGATGCGAATGCGGATGCGGATGCCGACTTTTGCATCGCCGAGCC	LAMP	Yasuhara et al. (2018)
Quencher probe	TCGGCATCCGCATCCGCATTCGCATCCGGGTCCTCAGCGT/3BHQ_1/	LAMP	Kubota et al. (2011)

3.4.2 Loop-Mediated Isothermal Amplification Assay

Loop-mediated isothermal amplification (LAMP) is a highly sensitive molecular technology that has been successfully employed in the detection of several plant pathogens (Sharma et al., 2017; Kashyap et al., 2016; Elkins et al., 2015; Villari et al., 2013; Notomi et al., 2000). Unlike several other nucleic acid–based PCR assays, LAMP reactions are isothermal. As a result, there is no requirement for expensive thermal cycling equipment. Besides requiring less reaction time not greater than 1 h, it is also less sensitive to PCR inhibitors, permitting the use of crude DNA extracts (Kogovšek et al., 2015; Tomlinson et al., 2007; Notomi et al., 2000). In addition to these advantages, it does not require extensive training for personnel performing the assay (Thiessen et al., 2016; Tomlinson et al., 2013;Lievens & Thomma, 2005), making this technology particularly suitable for field implementation by non-technical users. Moreover, successful amplifications can be visually observed by a turbidity change in the reaction tubes (Mori et al., 2001). Recently, LAMP has been optimised in the quantitative and specific detection of MoT fungus by Yasuhara-Bell et al. (2018). They developed LAMP primers by targeting PoT2 and MoT3 loci that have shown specificity towards *M. oryzae* and MoT, respectively (Table 3.2). The specificity of the primer set was determined and validated using 158 *M. oryzae* strains collected from different areas containing infected wheat and other grasses. The sensitivity was assessed using 10-fold serial dilutions of MoT gDNA with 50 *Fusarium* spp. isolates used as the negative control. PoT2- and MoT3-based assays showed high specificity for *M. oryzae* and MoT, respectively, and sensitivity was reported to be ~5 pg DNA reaction^{-1}. Using a field DNA extraction kit and the portable Genie II system, PoT2 and MoT3 assays tested on MoT-infected wheat seed and spikes identified *M. oryzae* and MoT, respectively. The mitochondrial *nad5* gene (NADH-dehydrogenase), an internal control for plant DNA, was multiplexed with PoT2 and MoT3 and showed comparable results to individual assays. The field implementation of the qLAMP assay system developed in this study has the potential to allow for accurate and timely detection of MoT spores before the development of visible symptoms, which is particularly useful for wheat blast pathogens that incite sporadic disease outbreaks or that can be easily confused with less damaging pathogens at the onset of the disease epidemic. Recently, Thierry et al. (2020) designed a set of markers that can be used in PCR, real-time PCR and LAMP for the detection WB fungus, with improved precision and specificity relative to earlier documented diagnostic assays. The major merit of these markers is their applicability on the isolates missing MoT3 sequence, for instance BR0032 isolate (Gladieux et al. 2018; Gupta et al. 2019; Yasuhara-Bell et al. 2019; Thierry et al. 2019). Interestingly, the developed LAMP assay allowed systematic detection of the WB fungus in seed lots containing at least two contaminated seeds or as low as 5 pg of DNA of fungus.

3.5 CONCLUSION AND FUTURE OUTLOOK

The exponential expansion of the MoT fungus in different countries is a serious threat to food and nutritional security for the ever-increasing population of the world. It has been realised that an MoT outbreak is not only a problem for one nation but it is also a global issue; moreover, geographical borders have no meaning for microbial pathogens. Therefore, continuous MoT monitoring is required to understand the genetics of the MoT population at the global level, including South Asian countries. It is critical to identify the pathogen below species level during the initial stages of infection, whether this is research driven or compelled by the need for rapid and accurate diagnostics during a disease outbreak. The availability of a detection system would be helpful in early decision-making to initiate management in order to avoid financial loss. Ultimately, the development of a convenient, portable, real-time and field-enabled molecular diagnostic tool is urgently needed for rapid monitoring and surveillance of wheat blast. Hence, for rapid mitigation of this new enemy of the wheat crop in South Asia, an intensive global effort is needed through practicing open data sharing and open science approaches for early and precise detection and diagnosis of wheat blast pathogens.

Recently, promising amplification platforms, probe development, various quantitative PCRs, DNA barcoding, RNA-Seq-based next-generation sequencing, nanotechnology and sensor technology have revolutionised the research field of fungal detection and diagnosis (Kashyap et al., 2016, 2017b, 2019; Sharma et al., 2017; Aslam et al., 2017). Although molecular diagnostics techniques have made great progress over the last decade (Donoso & Valenzuela, 2018), there is still a long way to go in the development and application of molecular diagnostics to assist field researchers in accurately diagnosing MoT. Additionally, molecular diagnostic procedures employed in plant disease diagnostic clinics need to be rapid, reliable, sensitive, inexpensive and easy to use and they need to compete with and complement traditional techniques to accurately pinpoint MoT infection in field-grown wheat and other weed hosts. Currently, a challenging task for crop researchers is to develop practical and field-applicable techniques for effective diagnostic assessment. In this direction, PCR-based technologies such as LAMP have been developed for MoT detection, but they may fail to distinguish viable from dead cells, as reported in the detection of several other microbes (Donoso & Valenzuela, 2018). This discrimination is vital when comparing plant samples post application of fungicides as DNA from dead cells can be detected and lead to erroneous interpretations and results. The use of propidium monoazide (PMA) could be an effective tactic to surmount this obstacle. PMA is a dye that selectively enters the cells' membranes and is effective in differentiating dead cells. After penetrating the cells, PMA covalently cross-links with DNA, resulting in a false PCR amplification (Nocker et al., 2007). Currently, PMA is employed in the detection of several microorganisms based on conventional PCR, real-time PCR and LAMP (Yan et al., 2017; Zhong et al., 2016; Pan & Breidt, 2007). PMA-based diagnostic procedures need effective sample tube incubation under halogen lamps (~650 W), making this methodology complex and tricky to employ with portable devices in automatic mode. However, easy-to-operate diagnosis platforms could be placed in farmers' wheat fields with a regular electricity supply.

Significant progress has been made in devising sensor-based technologies and their deployment for plant disease detection is quite impressive. New innovations in the area of precision agriculture and plant phenotyping have provided a toolbox of prospective solutions for crop health and stress management (Mahlein, 2016; Kumar & Kashyap, 2014; Cobb et al., 2013; Steddom et al., 2005). Currently, sensors are widely and successfully employed for non-invasive appraisal of crop nutrition status in farming land (Martinelli et al., 2014). The wide range and availability of low-cost sensors with adequate performance in the market is another important development for practical applications in agriculture (Wang et al., 2016; Paulus et al., 2014; Kashyap et al., 2017a). Unfortunately, specific and sensitive sensors for the detection of plant diseases are not available at the present time. Moreover, for effective and complete utilisation of potential sensors for disease detection have still need to be improvised. However, precise and effective instruments with revolutionary technological innovations can be tailored for specific and precise diagnosis and monitoring of wheat blast infection under field conditions will be available in the near future.

In conclusion, the technologies described in this chapter have a promising future in concert with the rapid advances in nanotechnology and optical and android phone–based sensors that will enable the production of cheaper, more precise and wireless systems for quick and real-time detection and diagnosis of crop diseases. As the MoT fungus is present in different wheat-growing areas, it can also be expected that these technological innovations will be exploited for automated in-field stations that can detect the airborne inocula of the MoT fungus and make data available for modelling and predicting WB epidemics well in advance.

REFERENCES

Anjos, J. R. N. and M. J. A. Charchar. "Natural Infection of Barley by *Pyricularia grisea* in Brazil." *Fitopatologia Brasileira* 25 (2000): 205.

Aslam, S., A. Tahir, M. F. Aslam, M. W. Alam, A. A. Shedayi and S. Sadia. "Recent Advances in Molecular Techniques for the Identification of Phytopathogenic fungi – A Mini Review." *Journal of Plant Interactions* 12, no. 1 (2017): 493–504. doi:10.1080/17429145.2017.1397205.

Bockus, W. W., C. D. Cruz, J. P. Stack and B. Valent. "Effect of Seed-Treatment Fungicides on Sporulation of *Magnaporthe oryzae* from Wheat Seed." *Plant Disease Management Reports* 9 (2015): ST004.

Callaway, E. "Devastating Wheat Fungus Appears in Asia for the First Time." *Nature* 532 (2016): 421–2.

Cardoso, C. D. A., E. M. Reis and E. N. Moreira. "Development of a Warning System for Wheat Blast Caused by *Pyricularia grisea*." *Summa Phytopathologica* 34 (2008): 216–21.

Castroagudín, V. L., P. C. Ceresini, S. C. de Oliveira and J. Teodora de Assis Reges. "Resistance to QoI Fungicides is Widespread in Brazilian Populations of the Wheat Blast Pathogen *Magnaporthe oryzae*." *Phytopathology* 105 (2015): 284–94.

Ceresini, P. C., V. L. Castroagudín, F. Á. Rodrigues et al. "Wheat Blast: Past, Present, and Future." *Annual Review of Phytopathology* 56 (2018): 427–56.

Chakdar, H., S. K. Goswami, E. Singh et al. "*noxB*-Based Marker for *Alternaria* spp.: A New Diagnostic Marker for Specific and Early Detection in Crop Plants." *3 Biotech* 9, no. 7 (2019): 249.

Cobb, J. N., G. DeClerck, A. Greenberg, R. Clark and S. McCouch. "Next-Generation Phenotyping: Requirements and Strategies for Enhancing Our Understanding of Genotype-Phenotype Relationships and its Relevance to Crop Improvement." *Theoretical and Applied Genetics* 126 (2013): 867–87.

Couch, B. C. and L. M. Kohn. "A Multilocus Gene Genealogy Concordant with Host Preference Indicates Segregation of a New Species, *Magnaporthe oryzae*, from *M. grisea*." *Mycologia* 94 (2002): 683–93.

Cruz, C. D. and B. Valent. "Wheat Blast Disease: Danger on the Move." *Tropical Plant Pathology* 42, no. 3 (2017): 210–22.

Cruz, C. D., W. W. Bockus, J. P. Stack et al. "Preliminary Assessment of Resistance Among U.S. Wheat Cultivars to the *Triticum* Pathotype of *Magnaporthe oryzae*." *Plant Disease* 96 (2012): 1501–5.

Cruz, C. D., J. Kiyuna, W. W. Bockus, T. C. Todd, J. P. Stack and B. Valent. "*Magnaporthe oryzae* Conidia on Basal Wheat Leaves as a Potential Source of Wheat Blast Inoculum." *Plant Pathology* 64 (2015): 1491–8.

Donoso, A. and S. Valenzuela. "In-Field Molecular Diagnosis of Plant Pathogens: Recent Trends and Future Perspectives." *Plant Pathology* 67, no. 7 (2018): 1451–61.

Duveiller, E., P. K. Singh, M. Mezzalama, R. P. Singh and A. Dababat. *Wheat Diseases and Pests: A Guide for Field Identification*. 2nd ed. Mexico: CIMMYT, Mexico, D.F., 2012.

Elkins, R. B., T. N. Temple, C. A. Shaffer et al. "Evaluation of Dormant-Stage Inoculum Sanitation as a Component of a Fire Blight Management Program for Fresh Market Bartlett Pear." *Plant Disease* 99 (2015): 1147–52.

Farman, M. L. "*Pyricularia grisea* Isolates Causing Gray Leaf Spot on Perennial Ryegrass (*Lolium perenne*) in the United States: Relationship to *P. grisea* Isolates from Other Host Plants." *Phytopathology* 92 (2002): 245–54.

Farman, M., G. Peterson, L. Chen et al. "The *Lolium* Pathotype of *Magnaporthe oryzae* Recovered from a Single Blasted Wheat Plant in the United States." *Plant Disease* 101 (2017): 684–92.

Gladieux, P., B. Condon, S. Ravel et al. "Gene Flow Between Divergent Cereal- and Grass-Specific Lineages of the Rice Blast Fungus *Magnaporthe oryzae*." *mBio* 9 (2018): e01219-01217.

Goulart, A. C. P. and F. A. Paiva. "Transmission of *Pyricularia oryzae* by Wheat *Triticum aestivum* Seeds." *Fitopatologia Brasileira* 15 (1990): 359–62.

Goulart, A. C. P., P. G. Sousa and A. S. Urashima. "Damages in Wheat Caused by Infection of *Pyricularia grisea*." (In Portuguese.) *Summa Phytopathologica* 33 (2007): 358–63. doi:10.1590/S0100-54052007000400007.

Gupta, D. R., C. S. R. Avila, J. Win et al. "Cautionary Notes on Use of the MoT3 Diagnostic Assay for *Magnaporthe oryzae* Wheat and Rice Blast Isolates." *Phytopathology* 109, no. 4 (2019): 504–8. doi:10.1094/PHYTO-06-18-0199-LE.

Hayashida, K., K. Kajino, L. Hachaambwa, B. Namangala and C. Sugimoto. "Direct Blood Dry LAMP: A Rapid, Stable, and Easy Diagnostic Tool for Human African Trypanosomiasis." *PLoS Neglected Tropical Diseases* 9 (2015): e0003578.

Igarashi, S. "Update on Wheat Blast (*Pyricularia oryzae*) in Brazil." In *Wheat for the Nontraditional Warm Areas: A Proceedings of the International Conference*, edited by D. A. Saunders, 480–3. Brazil: Foz do Iguaçu, 1990.

Igarashi, S., C. M. Utiamada, L. C. Igarashi, A. H. Kazuma and R. S. Lopes. "*Pyricularia* em trigo. 1. Ocorrência de *Pyricularia* sp. no estado do Paraná." *Fitopatologia Brasileira* 11 (1986): 351–2.

Islam, M. T., D. Croll, P. Gladieux et al. "Emergence of Wheat Blast in Bangladesh was Caused by a South American Lineage of *Magnaporthe oryzae*." *BMC Biology* 14 (2016): 11.

Kashyap, P. L., S. Kumar and A. K. Srivastava. "Nanodiagnostics for Plant Pathogens." *Environmental Chemistry Letters* 15, no. 1 (2017a): 7–13.

Kashyap, P. L., P. Rai, S. Sharma et al. "Nanotechnology for the Detection and Diagnosis of Plant Pathogens." In *Nanoscience in Food and Agriculture 2, Sustainable Agriculture Reviews 21*, edited by S. Ranjan et al., 253–76. Basel: Springer, 2016. doi:10.1007/978-3-319-39306-3_8.

Kashyap, P. L., P. Jasrotia, S. Kumar, D. P. Singh and G. P. Singh. "Identification Guide for Major Diseases and Insect-Pests of Wheat." *Technical Bulletin No. 18*, 17–18, 2018.

Kashyap, P. L., S. Kaur, G. S. Sanghera, S. S. Kang and P. P. S. Pannu. "Novel Methods for Quarantine Detection of Karnal Bunt (*Tilletia indica*) of Wheat." *Elixir Agriculture* 31 (2011): 1873–6.

Kashyap, P. L., S. Kumar, P. Jasrotia, D. P. Singh and G. P. Singh. "Nanosensors for Plant Disease Diagnosis: Current Understanding and Future Perspectives." In *Nanoscience for Sustainable Agriculture*, edited by R. Pudake, N. Chauhan and C. Kole, 189–205. Cham: Springer, 2019. doi:10.1007/978-3-319-97852-9_9.

Kashyap, P. L., P. Rai, S. Kumar, H. Chakdar and A. K. Srivastava. "DNA Barcoding for Diagnosis and Monitoring of Fungal Plant Pathogens." In *Molecular Markers in Mycology. Fungal Biology*, edited by B. Singh and V. Gupta, 87–122. Cham: Springer, 2017b.

Kogovšek, P., J. Hodgetts, J. Hall et al. "LAMP Assay and Rapid Sample Preparation Method for On-Site Detection of Flavescence Dorée Phytoplasma in Grapevine." *Plant Pathology* 64 (2015): 286–96.

Kohli, M. M., Y. R. Mehta, E. Guzman, L. de Viedma and L. E. Cubilla. "*Pyricularia* Blast – A Threat to Wheat Cultivation." *Czech Journal of Genetics and Plant Breeding* 47 (2011): S130–4.

Kubota, R., A. M. Alvarez, W. W. Su and D. M. Jenkins. "FRET-419 Based Assimilating Probe for Sequence-Specific Real-Time Monitoring of Loop-Mediated Isothermal Amplification (LAMP)." *Biological Engineering Transactions* 4 (2011): 81–100.

Kumar, S. and P. L. Kashyap. "Expanding Horizons of Precision Farming Driven Crop Protection." In *Precision Farming: A New Approach*, edited by Tulasa Ram, Shiv Kumar Lohan, Ranveer Singh and Purshotam Singh, 192–212. New Delhi: Daya Publishing House, 2014.

Lang, J. M., J. P. Hamilton, J. G. Q. Diaz et al. "Genomics-Based Diagnostic Marker Development for *Xanthomonas oryzae* pv. *oryzae* and *X. oryzae* pv. *oryzicola*." *Plant Disease* 94 (2010): 311–19.

Lang, J. M., P. Langlois, M. H. R. Nguyen et al. "Sensitive Detection of *Xanthomonas oryzae* Pathovars Oryzae and *oryzicola* by Loop-Mediated Isothermal Amplification." *Applied and Environmental Microbiology* 80 (2014): 4519–30.

Lau, D., F. M. Santana, J. L. N. Maciel et al. "Doenças de trigo no Brasil." In *Trigo no Brasil: bases para produção competitiva e sustentável*, edited by J. L. F. Pires, L. Vargas and G. R. da Cunha, 283–324. Passo Fundo: Embrapa Trigo, 2011.

Lievens, B. and B. P. H. J. Thomma. "Recent Developments in Pathogen Detection Arrays: Implications for Fungal Plant Pathogens and Use in Practice." *Phytopathology* 95 (2005): 1374–80.

Maciel, J. L. N., P. C. Ceresini, V. L. Castroagudin, M. Zala, G. H. J. Kema and B. A. Mcdonald "Population Structure and Pathotype Diversity of the Wheat Blast Pathogen *Magnaporthe oryzae* 25 Years After its Emergence in Brazil." *Phytopathology* 104 (2014): 95–107.

Mahlein, A. K. "Plant Disease Detection by Imaging Sensors – Parallels and Specific Demands for Precision Agriculture and Plant Phenotyping." *Plant Disease* 100, no. 2 (2016): 214–51.

Malaker, P. K., N. C. D. Barma, T. P. Tiwari et al. "First Report of Wheat Blast Caused by *Magnaporthe oryzae* Pathotype *Triticum* in Bangladesh." *Plant Disease* 100 (2016): 2330.

Marangoni, M. S., M. P. Nunes, N. Jr.Fonseca and Y. R. Mehta "*Pyricularia* Blast on White Oats: A New Threat to Wheat Cultivation." *Tropical Plant Pathology* 38 (2013): 198–202.

Martinelli, F., R. Scalenghe, S. Davino et al. "Advanced Methods for Plant Disease Detection. A Review." *Agronomy for Sustainable Development* 35 (2014): 1–25.

Martínez, S. I., A. Sanabria, M. C. Fleitas, V. F. Consolo and A. Perelló. "Wheat Blast: Aggressiveness of Isolates of *Pyricularia oryzae* and Effect on Grain Quality." *Journal of King Saud University – Science* (2018). doi:10.1016/j.jksus.2018.05.003.

Mehta, Y. R. and A. Baier. "Variação patogênica entre isolados de *Magnaporthe grisea* atacando triticale e trigo no estado do Paraná." *Summa Phytopathologica* 24 (1998): 119–25.

Mehta, Y. R., M. P. Nunes and J. C. Oliveira. "Occurrência de brusone em aveia no Estado do Paraná." In *Resultados Experimentais. XXVI Reunião da Comissão Brasileira de Pesquisa de Aveia*, 4–6 de abril, 55–7. Guarapuava, Paraná: FAPA, 2006.

Monsur, M. A., M. Ahmed, A. Haque et al. "Cross Infection Between Rice and Wheat Blast Pathogen *Pyricularia oryzae*." *Bangladesh Rice Journal* 20, no. 2 (2016): 21–9.

Mori, Y., K. Nagamine, N. Tomita and T. Notomi "Detection of Loop-Mediated Isothermal Amplification Reaction by Turbidity Derived from Magnesium Pyrophosphate Formation." *Biochemical and Biophysical Research Communications* 289 (2001): 150–4.

Nocker, A., P. Sossa-Fernandez, M. D. Burr and A. K. Camper. "Use of Propidium Monoazide for Live/Dead Distinction in Microbial Ecology." *Applied and Environmental Microbiology* 7 (2007): 5111–7.

Notomi, T., H. Okayama, H. Masubuchi et al. "Loop-Mediated Isothermal Amplification of DNA." *Nucleic Acids Research* 28 (2000): e63.

Pagani, A. P. S., A. C. Dianese and A. C. Café-Filho. "Management of Wheat Blast with Synthetic Fungicides, Partial Resistance and Silicate and Phosphite Minerals." *Phytoparasitica* 42 (2014): 609–17.

Pan, Y. and F. Breidt. "Enumeration of Viable Listeria Monocytogenes Cells by Real-Time PCR with Propidium Monoazide and Ethidium Monoazide in the Presence of Dead Cells." *Applied and Environmental Microbiology* 73 (2007), 8028–31.

Paulus, S., J. Behmann, A. K. Mahlein, L. Plümer and H. Kuhlmann. "Low-Cost 3D Systems: Suitable Tools for Plant Phenotyping." *Sensors* 14 (2014): 3001–18.

Pieck, M. L., A. Ruck, M. L. Farman et al. "Genomics-Based Marker Discovery and Diagnostic Assay Development for Wheat Blast." *Plant Disease* 101, no. 1 (2017): 103–9.

Rios, J. A., D. Debona, H. S. S. Duarte and F. A. Rodrigues. "Development and Validation of a Standard Area Diagram set to Assess Blast Severity on Wheat Leaves." *European Journal of Plant Pathology* 136 (2013): 603–11.

Sadat, M. A. and J. Choi. "Wheat Blast: A New Fungal Inhabitant to Bangladesh Threatening World Wheat Production." *Plant Pathology Journal* 33, no. 2 (2017): 103–8.

Saharan, M. S., S. C. Bhardwaj, R. Chatrath, P. Sharma, A. K. Choudhary and R. K. Gupta "Wheat Blast Disease: An Overview." *Journal of Wheat Research* 8, no. 1 (2016): 1–5.

Sharma, S., P. Rai, S. Rai et al. "Genomic Revolution in Crop Disease Diagnosis: A Review." In *Plants and Microbes in an Ever Changing Environment*, edited by Satya Shila Singh, 257–93. Hauppauge, NY: Nova Science Publishers, 2017.

Singh, D. P. "Wheat Blast: A New Challenge to Wheat Production in South Asia." *Indian Phytopathology* 70, no. 2 (2017a): 169–77.

Singh, D. P. "Wheat Blast Caused by *Magnaporthe oryzae* Pathotype *Triticum*: Present Status, Variability, and Strategies for Management." In *Management of Wheat and Barley Diseases*, edited by D. P. Singh, 635–42. Waretown, NJ: Apple Academic Press, 2017b.

Steddom, K., M. W. Bredehoeft, M. Khan and C. M. Rush. "Comparison of Visual and Multispectral Radiometric Disease Evaluations of *Cercospora* Leaf Spot of Sugar Beet." *Plant Disease* 89 (2005): 153–8.

Thierry, M., P. Gladieux, E. Fournier, D. Tharreau and R. Ioos. "A Genomic Approach to Develop a New qPCR Test Enabling Detection of the *Pyricularia oryzae* Lineage Causing Wheat Blast." *Plant Disease* 104 (2019): 60–70.

Thierry, M., A. Chatet, E. Fournier, D. Tharreau and R. Ioos. "A PCR, qPCR, and LAMP Toolkit for the Detection of the Wheat Blast Pathogen in Seeds." *Plants* 9, no. 2 (2020): 277. doi.org/10.3390/plants9020277

Thiessen, L. D., J. A. Keune, T. M. Neill, W. W. Turechek, G. G. Grove and W. F. Mahaffee. "Development of a Grower-Conducted Inoculum Detection Assay for Management of Grape Powdery Mildew." *Plant Pathology* 65 (2016): 238–49.

Toledo, J. and R. Escobar. *Piricularia o bruzone del trigo.* Santa Cruz, CA: CIAT, 2002.

Tomlinson, J. A., I. Barker and N. Boonham. "Faster, Simpler, More-Specific Methods for Improved Molecular Detection of Phytophthora Ramorum in the Field." *Applied and Environmental Microbiology* 73 (2007): 4040–7.

Tomlinson, J. A., S. Ostoja-Starzewska and I. P. Adams. "Loop-Mediated Isothermal Amplification for Rapid Detection of the Causal Agents of Cassava Brown Streak Disease." *Journal of Virology Methods* 191 (2013): 148–54.

Urashima, A. S. and H. Kato. "Varietal Resistance and Chemical Control of Wheat Blast Fungus." *Summa Phytopathologica* 20 (1994): 107–12.

Urashima, A. S., C. Grosso, A. Stabili et al. "Effect of *Magnaporthe grisea* on Seed Germination, Yield and Quality of Wheat." In *Advances in Genetic, Genomics and Control of Rice Blast Disease*, edited by G. L. Wang and B. Valent, 267–77. Dordrecht: Springer, 2009.

Valent, B., W. Bockus, C. Cruz et al. "Recovery Plan for Wheat Blast Caused by *Magnaporthe oryzae Triticum* Pathotype." In *USDA National Plant Disease Recovery System*, 2013. Accessed March 15, 2019. http://www.ars.usda.gov/SP2UserFiles/Place/00000000/opmp/Wheat%20Blast%20Recovery%20Plan%20Final.pdf.

Valent, B., C. D. Cruz, M. Farman et al. "Strategies for Managing Blast Disease of Wheat." Abstracts. *Achieving Durable Resistance to Wheat Diseases and Pests*, 5. 2–3 November 2016, Bloomington, MN.

Villari, C., J. A. Tomlinson, A. Battisti, N. Boonham, P. Capretti and M. Faccoli. "Use of Loop-Mediated Isothermal Amplification for Detection of *Ophiostoma clavatum*, the Primary Blue Stain Fungus Associated with Ips Acuminatus." *Applied and Environmental Microbiology* 79 (2013): 2527–33.

Vincelli, P. and N. Tisserat. "Nucleic Acid–Based Pathogen Detection in Applied Plant Pathology." *Plant Disease* 92, no. 5 (2008): 660–9.

Wang, H., F. Qin, L. Ruan et al. "Identification and Severity Determination of Wheat Stripe Rust and Wheat Leaf Rust Based on Hyperspectral Data Acquired Using a Black-Paper-Based Measuring Method." *PLoS ONE* 11, no. 4 (2016): e0154648.

White, T. J., T. Bruns, S. Lee and J. Taylor. "Amplification and Direct Sequencing of Fungal Ribosomal RNA Genes for Phylogenetics." In *PCR Protocols: A Guide to Method and Amplifications*, edited by M. A. Innis, D. H. Gelfand, J. J. Sninksky and T. J. White, 315–22. San Diego, CA: Academic, 1990.

Wise, K., C. Woloshuk and A. Freije. *Diseases of Wheat: Fusarium Head Blight (Head Scab)*, BP-33-W. Purdue University Extension. (2015): 1–4.

Yan, M., L. Xu, H. Jiang, Z. Zhou, S. Zhou and L. Zhang. "PMA-LAMP for Rapid Detection of *Escherichia coli* and Shiga Toxins from Viable but Non-Culturable State." *Microbial Pathogenesis* 105 (2017): 245–50.

Yasuhara-Bell, J., K. F. Pedley, M. Farman, B. Valent and J. P. Stack. "Specific Detection of the Wheat Blast Pathogen (*Magnaporthe oryzae Triticum*) by Loop-Mediated Isothermal Amplification." *Plant Disease* 102 (2018): 2550–9.

Zhang, H. L., Z. S. Wu, C. F. Wang, Y. Li and J. R. Xu. "Germination and Infectivity of Microconidia in the Rice Blast Fungus *Magnaporthe oryzae*." *Nature Communications* 5 (2014): 4518.

Zhang, N., J. Luo, A. Y. Rossman et al. "Generic Names in Magnaporthales." *IMA Fungus* 7 (2016): 155–9.

Zhong, Q., J. Tian, B. Wang and L. Wang. "PMA Based Real-Time Fluorescent LAMP for Detection of *Vibrio Parahaemolyticus* in Viable but Nonculturable State." *Food Control* 63 (2016): 230–8.

4 Infection Process of *Magnaporthe oryzae* pathotype *Triticum* on Wheat

Fabrício Ávila Rodrigues, Caroline Hawerroth, Jonas Alberto Rios, Daniel Debona and Carlos Eduardo Aucique-Pérez

CONTENTS

4.1 INTRODUCTION

Blast caused by the hemibiotroph ascomycetous *Magnaporthe oryzae* pathotype *Triticum* (MoT) dramatically affects wheat yield and grain quality (Castroagudín et al., 2016; Ceresini et al., 2018; Kohli et al., 2011; Rodrigues et al., 2017; Rios et al., 2017). Host–pathogen interaction is a multifaceted process mediated by pathogen- and plant-derived molecules (Boyd et al., 2013). From the pathogen's side, the secreted effectors are determinants of the pathogenicity or aggressiveness towards the host. On the other hand, resistance proteins from the host recognise the pathogen's effectors and activate some defence mechanisms in a fast and orchestrated way to overcome the infection (Boyd et al., 2013). The strategies used by many species of fungal pathogens to infect their hosts are well known. A series of events must be completed by a particular fungal pathogen to infect its host successfully. The fungal pathogen needs to encounter its host; adhere to its surface; germinate and form the germ tube that will perceive the leaf surface topology; penetrate either directly (e.g. through the cuticle by forming a swollen dome-shaped cell differentiated from the end of the germ tube, called an appressorium) or indirectly (e.g. through stomata); efficiently colonise the tissues to obtain nutrients to support its continuous grow; counteract any defence mechanism by the host; and reproduce (Braun & Howard, 1994; Dean, 1997; Mendgen & Deising, 1993; Mendgen et al., 1996; Tucker & Talbot, 2001). By completing all these steps, the fungal pathogen is able to perpetuate its species in nature. The offsprings will then cause auto and/or alloinfections and keep the disease developing within the same crop season, building an inoculum to infect plants in the following season.

Considering the importance of blast in decreasing the rice yield worldwide, the rice–*Magnaporthe oryzae* interaction has been the focus of intense research and has become a model pathosystem to illustrate most of the concepts governing some host–pathogen interactions (Ebbole, 2007; Valent & Chumley, 1991; Veneault-Fourrey & Talbot, 2005; Wilson & Talbot, 2009). Unfortunately, aspects of the MoT infection process on the leaves and spikes of wheat plants are much less documented in contrast to the lessons learned from studies related to the infection process of *Magnaporthe oryzae* on rice (Howard, 1994; Howard & Valent, 1996; Rodrigues et al., 2003, 2005). The infection process of *M. oryzae* on rice (Howard, 1994; Howard & Valent, 1996; Rodrigues et al., 2003, 2005; Wilson & Talbot, 2009) and of MoT on wheat (Cruz et al., 2016) leaves shares some similarities especially regarding conidium adhesion and germination, appressorium formation and leaf tissue colonisation. On leaves, wheat blast symptoms start as grey-green and water-soaked lesions with dark green borders that become light tan with necrotic borders after they have fully expanded (Cruz et al., 2016). Lesions are grey-brown on spikelets and bright black on rachilla (Cruz et al., 2015b).

This chapter summarises our current understanding of the blast and MoT cycles on wheat and provides some details of the infection process of MoT on leaves and rachilla from a microscopic perspective.

4.2 BLAST CYCLE

Several aspects of the blast cycle on wheat need to be better elucidated, mainly the dynamics of the inoculum buildup for the occurrence of epidemics under variable environmental conditions (Alves & Fernandes, 2006; Cruz et al., 2015a). All aerial organs of wheat plants (e.g. culm, leaves and spikes) can be infected by MoT (Igarashi et al., 1986; Cruz et al., 2015b, 2016). The conidia of MoT produced on lesions that develop on the leaves of wheat plants are efficiently dispersed by rain-splash and wind and initiate fungal infections in the leaves or spikes of the same (autoinfections) or other (alloinfections) plants, increasing the blast epidemic rate (Figure 4.1). The conidia of MoT were reported to be dispersed up to 1 km from a group of artificially infected wheat plants (Urashima et al., 2007). On leaves, the lesions are elliptical or rounded with dark brown margins and a grey centre (Figure 4.1). Even though blast can negatively affect the photosynthesis of wheat leaves (Debona et al., 2012; Rios et al., 2013; Rodrigues et al., 2017), fungal infections on spikes originating from conidia produced on the foliar lesions have a great impact on a blast epidemic.

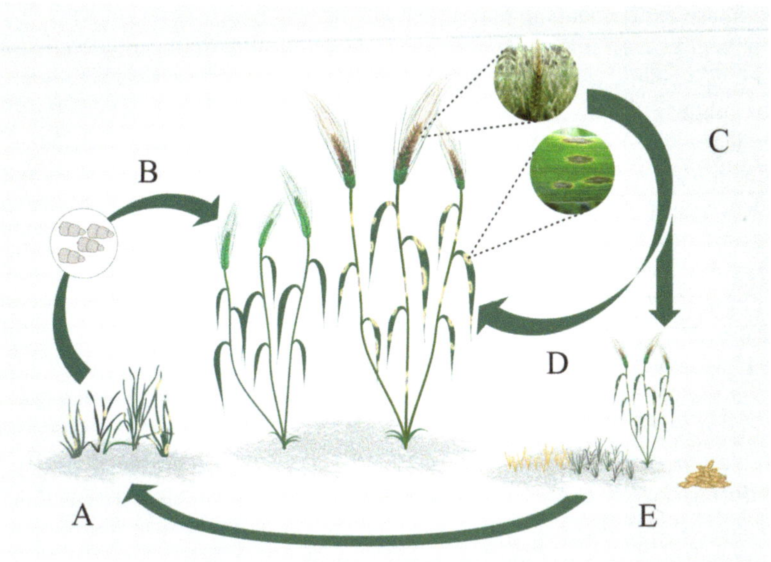

FIGURE 4.1 Blast cycle caused by *Magnaporthe oryzae* pathotype *Triticum* (MoT) on wheat plants. (A) Conidia of MoT are produced on the lesions formed on the leaves of wheat plants at the vegetative growth stage. (B) Conidia are then dispersed by rain-splash and wind and deposited on leaves and spikes to initiate fungal infections. (C) On leaves, blast lesions are elliptical or rounded with dark brown margins and with a grey centre; on spikelets, the lesions are grey-brown and cause premature bleaching of the spikes. (D) Auto and alloinfections by MoT increase the blast epidemic rate resulting in high disease severity on plants in the same growing season characterising a polycyclic disease. (E) MoT survives on wheat plants that remain in the field after harvest, on wheat residues in the soil and on seeds and is the initial inoculum for the occurrence of blast in the following season.

Blast symptoms on spikes and spikelets are very distinctive from those occurring on leaves (Debona et al., 2012; Rios et al., 2013). Lesions on spikes and spikelets are grey-brown, the rachis is bright black and the spikes can completely bleach out due to MoT infection on the rachis and peduncle (Figure 4.1) (Cruz et al., 2015b; Goulart et al., 1995; Urashima et al., 2009). Regarding spike blast, many aspects of its epidemiology need to be determined, including latent and incubation periods, optimum plant growth stage for fungal infection, microclimate and the interaction and dynamics of factors that support inoculum buildup and widespread epidemics (Cruz et al., 2015a; Rios et al., 2016).

MoT was found to be transmitted from infected wheat seeds to seedlings at rates ranging from 32% to 64% and resulted in intense necrosis of the coleoptiles and death of seedlings (Toledo, 2004). The blast epidemic rate is high when plants are exposed to cloudy days, temperatures ranging from 24°C to 28°C and relative humidity greater than 80% (Goulart et al., 2007). Temperatures ranging from 25°C to 30°C and a wetness duration greater than 25 hours favoured MoT infection on spikes and leaves (Cardoso et al., 2008). Cruz et al. (2015a) carried out field and greenhouse experiments to determine whether the basal senescent leaves of wheat plants could be an important source of inoculum of MoT to initiate infections on spikes. Blast severity was greater on older than younger leaves of wheat plants. In greenhouse conditions, 100% of wheat plants showed senescence on the lowest three leaves and conidia were found in 99% of them (Cruz et al., 2015a). Conidia production on infected leaves coincided with the emergence of spikes, but blast incidence on spikes was low. For a susceptible cultivar, a total of 1,669,000 conidia were obtained per gram of dried leaves from the lowest plant canopy (Cruz et al., 2015a). Across all cultivars tested, from 390,000 to 401,000 conidia were obtained per gram of basal senescent leaves (Cruz et al., 2015a). This amount of conidia is considered high compared to the amount needed (e.g. 24,000 conidia per spike) for high blast severity on plants of highly susceptible cultivars (Cruz et al., 2012). These findings indicate that attention needs to be paid to the lower canopy of wheat plants, especially those of susceptible cultivars, as an important source of inoculum for a high blast epidemic rate (Cruz et al., 2015a). MoT uses multiple strategies to survive during the wheat season. The fungus is able to infect wheat plants that remain after harvest and also alternative hosts, and to survive on wheat residues left in the soil and also on seeds (Figure 4.1). MoT is not able to survive freezing temperatures (Ebbole, 2007). Alternative hosts of MoT include barley, foxtail finger millet and millet as well as wild and cultivated grasses such as annual ryegrass, goosegrass, perennial ryegrass, St. Augustine grass and tall fescue (Prabhu et al., 1992; Urashima et al., 2004). A field experiment aiming to investigate the effect of blast on groups of wheat plants from different cultivars inoculated with MoT (0%, 5%, 10%, 20% and 30% of inoculated plants per plot at the completely emerged spikes growth stage) showed that blast incidence was greater than 50% for some cultivars when 5% of the plants were inoculated with MoT (Gomes et al., 2017). This finding indicates that a large amount of wheat seeds can potentially be infested or infected by MoT and serve as an efficient source of primary inoculum. The time of survival of MoT on seeds in the form of mycelium depends on the storage conditions. Wheat seeds with a MoT incidence ranging from 6% to 14% that were kept at room temperature showed a significant decrease in fungal survival after 6 months; only 1.5% of the seeds remained infected after 12 months (Goulart & Paiva, 1993). The viability of MoT was lost on seeds that originated from spikes naturally infected and stored in paper bags for 22 months (Reis et al., 1995). Considering that growers store wheat seeds for 7–8 months before sowing, infested or infected seeds become a potential source of primary inoculum to start a blast epidemic. Infection of the roots of wheat seedlings by MoT originating from infested and/or infected seeds results in conidia production that will infect the above part of the seedlings (Ebbole, 2007). The international trade in wheat seeds supposedly infested and/or infected by MoT plays a key role in long-distance fungal dispersal (Ceresini et al., 2018). Moreover, given the wide host range of MoT, it is even more difficult to establish the role played by wheat seeds as a major source of primary inoculum for blast epidemics. Considering that a significant amount of infected seeds of low weight are removed during harvesting or processing, this concern can be alleviated. However, the role of wheat seeds as the primary source of inoculum in areas where the blast is non-endemic cannot be overlooked considering the recent report of blast in Bangladesh (Ceresini et al., 2018; Islam et al., 2016).

4.3 THE LIFE CYCLE OF MoT

The events related to the life cycle of MoT on the leaves of wheat plants (Figure 4.2) share great similarities with the life cycle of *M. oryzae* on rice plants (Rodrigues et al., 2003, 2005) and are of detrimental importance for successful fungal infection on wheat plants. These events will be briefly discussed in the following subsections.

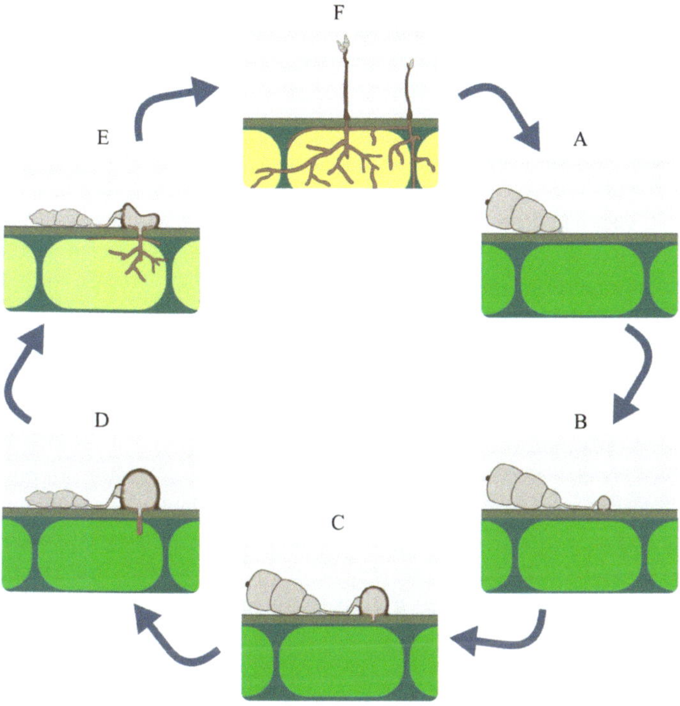

FIGURE 4.2 Life cycle of *Magnaporthe oryzae* pathotype *Triticum* (MoT) on the leaves of wheat plants. (A) Conidium of MoT is deposited over the leaf surface and initiates the adhesion process with the aid of the mucilage stored at its tip. (B) Conidium germinates producing a germ tube and a melanised appressorium at the tip of the germ tube in response to host cues. (C) The turgor pressure exercised by the melanised appressorium helps the penetration peg to penetrate directly into the cuticle and the epidermal cell wall. (D) During pre-penetration, the contents of the autophagocytised conidium and germ tube are transported to the inside of the melanised appressorium. (E) Infection hyphae of bulbous appearance that originated from the penetration peg ramify within the epidermal cell. (F) Ramified fungal hyphae within the first-penetrated epidermal cell colonise intra- and intercellularly the surrounding cells. At the advanced stage of fungal infection, MoT sporulates on the necrotic leaf tissue. Conidia are arranged in a sympodial fashion on the aerial conidiophores. Conidia produced in each lesion serve as the inoculum for the secondary fungal infection cycles.

4.3.1 Adhesion

Adhesion is the first step involved in the infection process of many fungal pathogens. Based on scanning electron microscopic observations, the conidium of MoT starts to germinate on the adaxial surface of wheat leaves 6 hours after inoculation (hai) (Figure 4.3A). Spore germination on the leaves of its host is always accompanied by the production of an extracellular matrix (ECM) (Xiao et al., 1994). The ECM formation is a key event for fungal infection prior to gaining full access to the host cell because it provides a firm base for penetration. Using immunological methods (antibodies against integrin and mammalian cell adhesion factors), Inoue et al. (2007) analysed the ECM components produced by MoT on wheat leaves and searched for enzymes that were able to digest the fungal ECM to better understand blast development. According to the authors, the ECM was characterised by fibrous and amorphous materials located in the spaces between the fungal cell walls and the plant cuticle. Immunohistochemical and immunoelectron microscopy suggested that ECM includes components that positively reacted with antibodies of four animal cell adhesion factors (collagen VI, vitronectin, fibronectin and laminin) and an animal integrin a3. Incubated on a cellulose membrane, the ECM was rapidly digested by matrix metalloproteinases (collagenase and gelatinase B), resulting in the detachment of most infection structures from the membrane surface. Both ultrastructural observations and immunological responses showed that more ECM was located at the appressoria than at the conidia and germ tubes. Moreover, the appressoria from MoT needed a powerful adhesion force for the aggressive action of penetration pegs into the cuticle on wheat leaves (Inoue et al., 2007). Fungal hydrophobins are cell-surface, hydrophobic proteins that are ubiquitous in filamentous fungi and found on the outer surface of conidia and hyphae, and aid in adhesion or communication between the fungus and its environment (Wösten, 2001). Hydrophobins are secreted to the cell surface and water atmosphere as monomer proteins and subsequently self-assemble. The self-assembled hydrophobins reduce the surface tension of the growth medium, allowing fungi to breach the water–air interface and develop aerial hyphae, permitting the efficient exchange of gas (Wang et al., 2005). Inoue et al. (2016) investigated the role of two hydrophobins, class I Mpg1 and class II Mhp1, on the adhesion and germling differentiation of MoT on wheat. Hydrophobin mutants were obtained using gene knockdown or knockout experiments. The Mpg1-silenced mutants had various levels of Mpg1 transcription, and Mpg1 expression was correlated with adhesion, germling differentiation and disease symptoms development. The authors also obtained mhp1 null mutants that had caused great blast symptoms on the leaves of both barley and wheat plants. Moreover, the Mpg1 directly contributed to adhesion, but the Mhp1 was not essential for the infection process of MoT on wheat leaves (Inoue et al., 2016).

4.3.2 Penetration

Appressoria formation plays a significant role in enabling some pathogens to penetrate their host's tissue (Dean, 1997; Mendgen & Deising, 1993). In this regard, the turgor pressure of a turgid appressorium is translated into a mechanical force

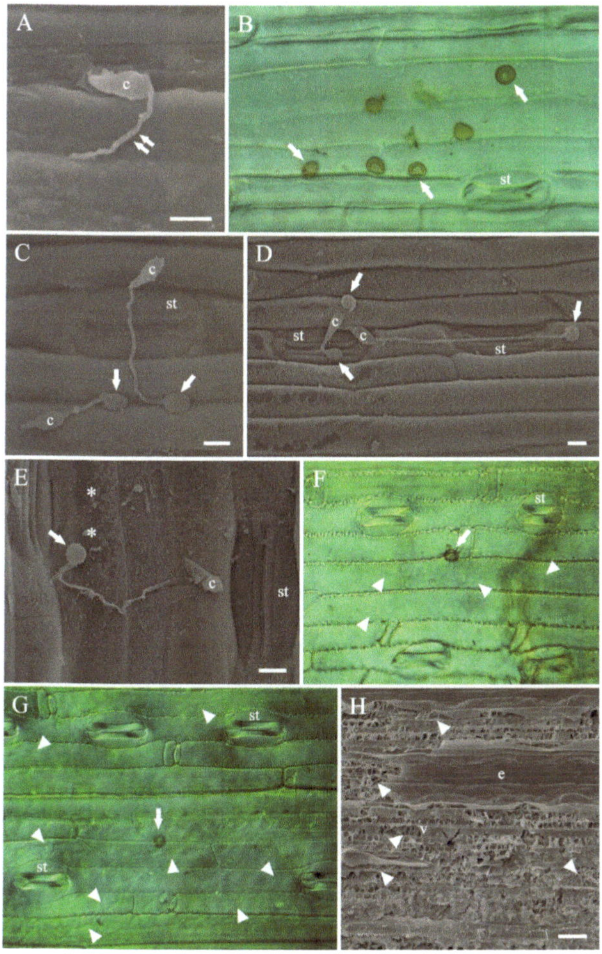

FIGURE 4.3 Scanning electron micrographs (A, C–E, H) and differential interference contrast microscopy (B, F, G) of the adaxial surface of wheat leaves inoculated with *Magnaporthe oryzae* pathotype *Triticum* (MoT). (A) At 6 hai, conidium germinated and formed the germ tube (double arrows). (B–D) Melanised appressoria (arrows) were formed at the tip of the germ tubes at 12 hai. (C) Germ tube originating from one conidium crossed the stomate without penetration to form a melanised appressorium (arrows) between the two epidermal cells. (D) The conidium formed two short germ tubes with two melanised appressoria (arrows). The conidium produced a long germ tube that crossed the stomate to form a melanised appressorium (arrow) over the leaf surface. (E) From 36 to 48 hai, the conidium and the germ tube showed signs of autophagy at the appressorial site (arrow). The epicuticular wax was degraded (asterisks). (F) At 48 hai, some epidermal cells were colonised by the hyphae of MoT (arrowheads) after fungal penetration from an appressorium (arrow). (G) At 72 hai, fungal hyphae massively colonised the epidermal cells (arrowheads) after fungal penetration from a melanised appressorium (arrow). (H) After cuticle removal, abundant fungal hyphae were noticed colonising the mesophyll (arrowheads) and some vascular vessels at 72 hai. Conidium (c), epidermis (e), stomate (st) and vascular vessel (v). Scale bars: A, C–E = 10 μm; B, F, G = 400×; and H = 50 μm.

that is exerted by the emerging penetration peg to allow it to penetrate the leaf cuticle (Dean, 1997; Mendgen & Deising, 1993). Following conidia germination, MoT needs to form an apressorium to guarantee its ingress into the wheat leaf tissue. The appressorium of MoT forms on wheat leaves as a dome-shaped cell with a complex cell wall structure at the tip of the germ tube after 12 hai (Figure 4.3B–D). Inoue et al. (2007) reported that the conidia of MoT were able to germinate and produce infection structures on cellulose membranes and on wheat leaves within 9 hai. During differentiation, appressoria usually started to produce a small amount of melanin in the spaces between the plasma membranes and the cell walls by 6 hai, accumulating a large amount of melanin by 9 hai (Inoue et al., 2007). During germination, the conidium of MoT can form two germ tubes (Figure 4.3D). It forms a long germ tube that crosses the stomate to form an appressorium over the leaf surface (Figure 4.3C and D). From 36 to 48 hai, the epicuticular wax on the adaxial leaf surface is disintegrated and both the conidium and germ tube show signs of autophagy (Figure 4.3E). Fungal pathogens secrete an array of hydrolytic enzymes to degrade the components of their host's epidermis, which include waxes, cuticle and cell walls. Cutinases and penetration pegs are necessary to help the fungus breach the host cuticle and the cell wall of the epidermal cell (Dean, 1997; Ribot et al., 2008; Wilson & Talbot, 2009). In rice, for example, the removal of the wax on the leaf surface by non-specific esterases released by *M. oryzae* might increase degradation of cuticle (Dean, 1997; Ribot et al., 2008; Wilson & Talbot, 2009). The external infection structures of fungal pathogens, such as conidia and appressoria, tend to languish after the fungus gains full access to the first-penetrated epidermal cell (Mendgen & Deising, 1993; Wilson & Talbot, 2009). Van Vu et al. (2012) provided RNAi-based evidence that cellulases belonging to glycosyl hydrolase (GH) families 6 and 7 were important for MoT to penetrate the epidermis of wheat leaves and colonise the leaf tissues. The GH6 and GH7 cellulases likely include all members of the cellobiohydrolase family and some endoglucanases in MoT. Quantitative reverse-transcriptase polymerase chain reaction analysis indicated that more than half of the cellulases were highly induced during the MoT infection process. The authors constructed knock-down (KD) mutants of these cellulases using the building blocks method. The transcript levels of the target genes and cellulase activity were considerably reduced in the KD mutants. The KD mutants resulted in fewer blast lesions, less fungal penetration and reduced colonisation of the leaf tissue in comparison to the wild strain. Cytological analyses revealed that a high number of papilla formed on the infection sites helped to block the invasion of the KD mutants into the leaf cells (Van Vu et al., 2012). Inoue et al. (2011) evaluated the role played by various enzymes produced on the germlings of MoT that could contribute to blast development on wheat leaves. According to the authors, treatment with β-mannosidase, collagenase N-2, collagenase S-1 or gelatinase B at 1 hai resulted in the detachment of germlings, although normal appressoria were formed. Treatment with matrix metalloproteinases (MMPs) at 6 hai also caused the detachment of germlings. Furthermore, using inoculation tests and scanning electron microscopy, the authors confirmed that the germlings on wheat leaves were removed and were unable to infect the wheat leaves receiving MMPs. In a pathogenicity test on wheat leaves, the formation of blast lesions was significantly reduced by treatment with trypsin despite a weak detachment ability.

This suppressive effect seemed to be involved in the inhibitory effects of invasion or other developmental stages rather than in the degradation of adhesion components. The absence of blast lesions occurred on leaves that received pronase E and some MMPs. The most effective MMPs were crude collagenase, collagenase S-1 and gelatinase B, suggesting that the application of MMPs may become a strategy to control blast on wheat because of their capacity to detach the germlings of MoT. Autophagy of the conidium and germ tube is essential during the formation of the functional appressorium because their contents are a source of carbon that helps to generate the turgor pressure needed for the penetration peg to penetrate the cuticle and the epidermal cell wall (Ebbole, 2007; Veneault-Fourrey et al., 2006).

4.3.3 Colonisation

In general, once inside the epidermal cell, the infection hyphae formed from the penetration peg of the fungal pathogens ramify intra- and intercellularly to gain access to nutrients and achieve intense leaf tissue colonisation (Mendgen & Deising, 1993; Wilson & Talbot, 2009). By 48 hai, MoT successfully forms an extensively branched mycelium in the first-penetrated epidermal cell and colonises a few neighbouring cells (Figure 4.3F). By 72 hai, fungal hyphae colonise many other epidermal cells (Figure 4.3G), the mesophyll cells and some vascular vessels (Figure 4.3H). At 80 hai, many conidiophores emerge from the adaxial surface of the disintegrated leaf tissue (Figure 4.4A). From 96 hai, conidiophores start to bear conidia (Figures 4.4B–E and 4.5A–D). Some conidia launched from the conidiophores germinate on the adaxial leaf surface to initiate autoinfections (Figure 4.4C and E). From 120 to 144 hai, MoT grows abundantly over the surface of the disintegrated leaf tissue (Figure 4.5A–D). The release of many non-host selective toxins during *M. oryzae* infection causes intense leaf tissue necrosis and nutrients available help the fungus to achieve intensive sporulation (Valent & Chumley, 1991; Rodrigues et al., 2003, 2005). In rice and wheat, the conidia formed on aerial conidiophores are important in the secondary cycle of *M. oryzae* (Cruz et al., 2015b, 2016; Ribot et al., 2008; Rodrigues et al., 2003, 2005; Wilson & Talbot, 2009). In rice, *M. oryzae* starts to sporulate on the necrotic lesions on rice leaves between 6 and 15 days after inoculation (Ribot et al., 2008) and on wheat leaves it starts at 5 days after inoculation (Rios et al., 2013; Debona et al., 2012; Cruz et al., 2016; Rodrigues et al., 2017).

On rachis tissue, the epidermis and parenchyma are colonised by the hyphae of MoT at 72 hai (Figure 4.6A and B). From 96 to 168 hai, the vascular vessels and the parenchyma are massively colonised by the hyphae of MoT (Figure 4.6C and D). Indeed, fungal hyphae grow both intra- and intercellularly in the vascular bundles (xylem vessels, phloem sieves) and the cortical and pith parenchyma (Figure 4.6E and F). Regarding the colonisation of rachis nodes by MoT, Cruz et al. (2015b) reported the presence of fungal hyphae in the epidermis, vascular bundles and cortical tissue at 96 hai. In the vascular bundles, fungal hyphae colonised both vertically and horizontally into the xylem vessels, phloem sieves and cortical and pith parenchyma cells. In the cortical parenchyma tissue outside the vascular bundles, fungal hyphae grew both inter- and intracellularly and were able to cross the host cell wall (Cruz et al., 2015b). Massive wheat yield losses due to blast occur when *M. oryzae*

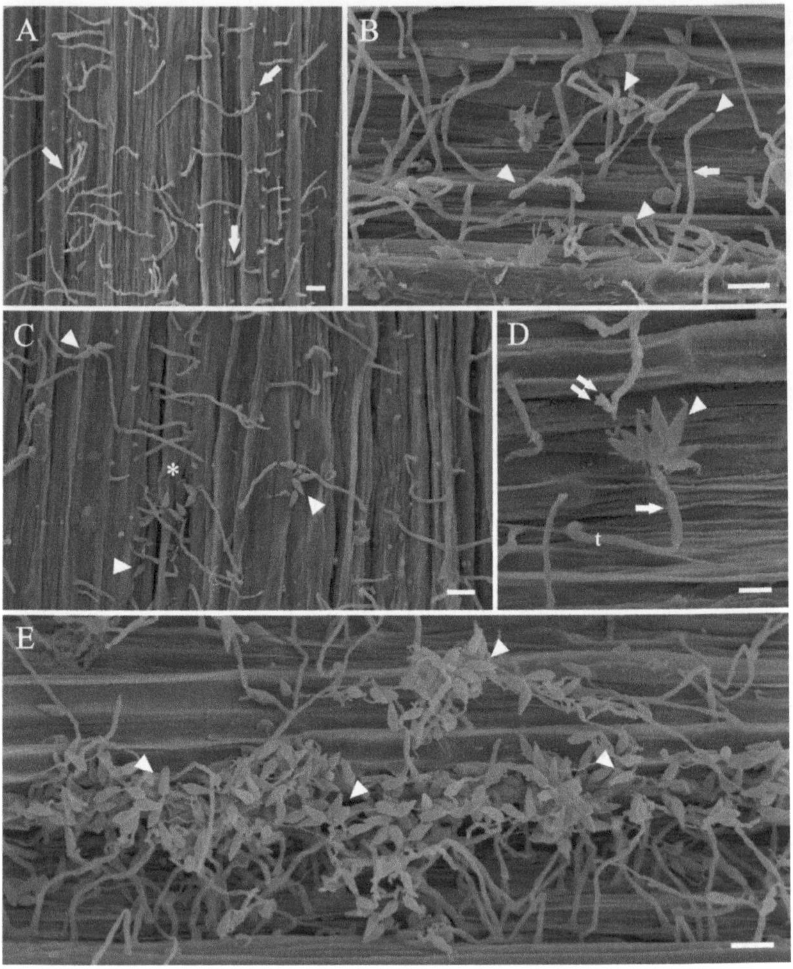

FIGURE 4.4 Scanning electron micrographs of the adaxial surface of wheat leaves inoculated with *Magnaporthe oryzae* pathotype *Triticum* (MoT). (A) At 80 hai, many conidiophores (arrows) emerged from the adaxial surface of the disintegrated leaf tissue. (B, C) At 96 hai, conidia were produced (arrowheads) on conidiophores (arrow). Conidia released from the conidiophores were deposited over the leaf surface, and some germinated (asterisk). (D) Detail of a conidiophore (arrow) bearing some conidia (arrowhead) and of scars on the conidiophore (double arrows) (96 hai). From 96 to 120 hai, many conidia were produced (arrowheads) on the conidiophores that emerged from the degraded leaf tissue. Trichome (t). Scale bars: A–C, E = 20 μm; D = 10 μm.

infects the rachis at the base of the spike, thereby limiting grain development and filling, in addition to completely destroying the spikes (Goulart et al., 2007; Kohli et al., 2011). The rachis node region is rich in vascular bundles, which are responsible for nutrient and water translocation to grains to guarantee better filling. A more pronounced colonisation of *M. oryzae* in the rachis, possibly due to the abundance

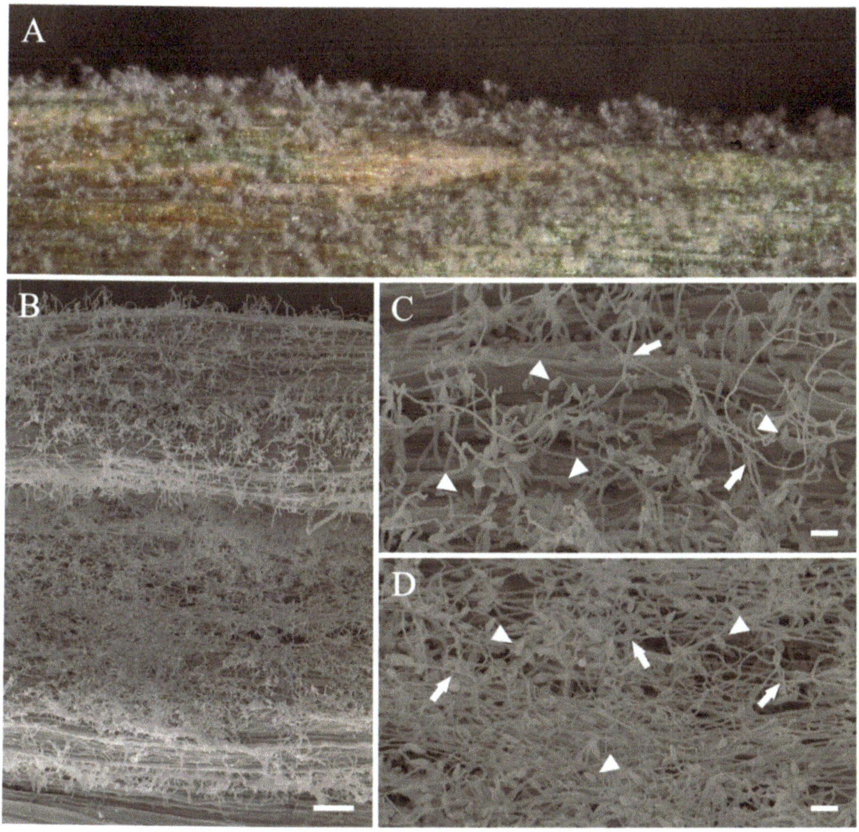

FIGURE 4.5 Stereomicroscopy image (A) and scanning electron micrographs (B–D) of the adaxial surface of wheat leaves inoculated with *Magnaporthe oryzae* pathotype *Triticum* (MoT). (A–D) From 96 to 120 hai, MoT grew abundantly over the leaf tissue as noticed by the massive mycelia (arrows) and conidiophores bearing conidia (arrowheads). Scale bars: A = 40×; B = 100 μm; C, D = 20 μm.

of nutrients, can greatly compromise grain filling considering the pivotal role of the rachis in wheat yield.

4.3.4 Other Aspects Related to the Infection Process of MoT on Wheat

Araujo et al. (2016) investigated the cytological aspects of incompatible (non-host resistance) and compatible (host resistance) rice–(R-*Mo*) and wheat–(W-MoT) *M. oryzae* and MoT isolates interactions, respectively. Inoculations of rice and wheat with the R-*Mo* and W-MoT fungal isolates, respectively, were compatible interactions (host resistance), whereas inoculations of rice and wheat with the W-MoT and R-*Mo* fungal isolates, respectively, were considered incompatible interactions (non-host resistance). For compatible interactions (rice–R-*Mo* and wheat–W-MoT), fungal hyphae penetrated and colonised the epidermal cells and also invaded many

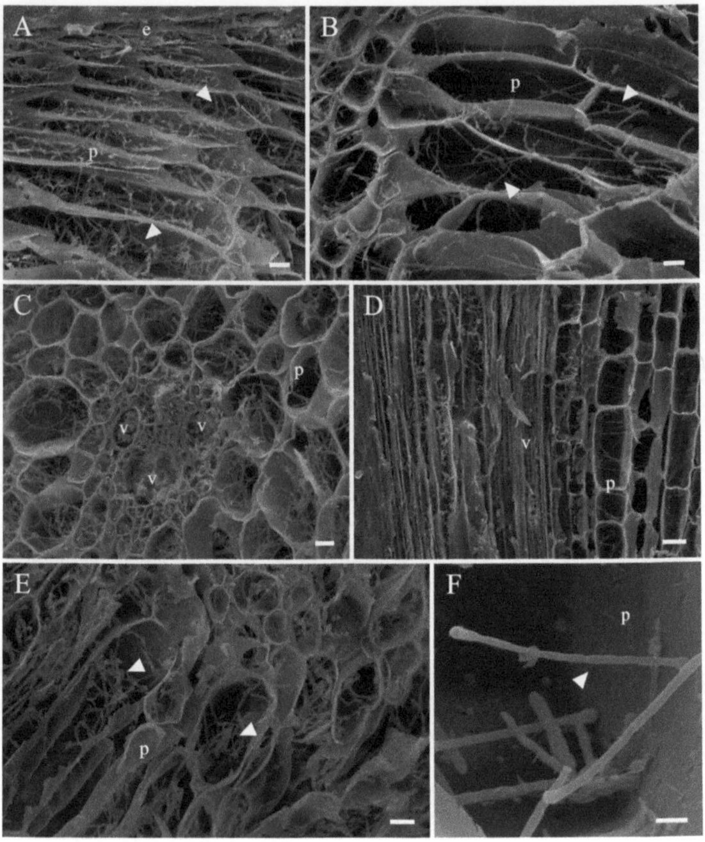

FIGURE 4.6 Scanning electron micrographs of the longitudinal (A, D–F) and transverse (B, C) sections of the rachis from wheat spikes inoculated with *Magnaporthe oryzae* pathotype *Triticum* (MoT). (A, B) Fungal hyphae (arrowheads) colonised the epidermal cells and reached the parenchyma at 72 hai. (C–E) From 96 to 168 hai, the parenchyma and the vascular vessels (arrowheads) were massively colonised by fungal hyphae. (F) Detail of fungal hyphae (arrowhead) inside a parenchyma cell at 168 hai. Epidermis (e), parenchyma (p) and vascular vessels (v). Scale bars: A–C, E = 20 μm; D = 50 μm; F = 5 μm.

neighbouring cells. By contrast, in the case of incompatible interactions (rice–W-MoT and wheat–R-*Mo*), fungal hyphae were not able to penetrate or colonise the epidermal cells, but when penetration did occur, the hyphae were restricted to the first-invaded epidermal cell. The frequency of appressorial sites exhibiting infection hyphae within the epidermal cell underlying an appressorium was greater in the case of compatible interactions. By contrast, unsuccessful penetrations with cytoplasmic granulation occurred with high frequency in incompatible wheat–R-*Mo* and rice–W-MoT interactions and the number of necrotic epidermal cells underlying the appressorium was low for the rice–W-MoT interaction as well as for the wheat–R-*Mo* interaction, where no symptoms of necrosis were exhibited. However, the opposite was observed for compatible interactions. Tufan et al. (2009) investigated the response of wheat to both adapted and non-adapted isolates of MoT at the cellular and transcriptional levels. Resistance to the non-adapted isolate of MoT was associated with the formation of appositions (halos) beneath attempted fungal penetration sites that wheat-adapted MoT isolates were able to breach. Indeed, transcriptome analysis indicated extensive transcriptional reprogramming following inoculation with both wheat-adapted and non-adapted isolates of MoT. The functional annotation of many of the differentially expressed transcripts was classified into categories: cell rescue and defence; plant metabolism; cellular transport; and regulation of transcription. The authors also reported that defence-related transcripts commonly induced by adapted and non-adapted isolates were differentially regulated in response to MoT and *M. oryzae* isolates over time. The differential expression of genes involved in cellular transport indicated the importance of this process in wheat defence to blast (Tufan et al., 2009). According to Tufan et al. (2012), members of the *Wheat-Induced Resistance 1* (*TaWIR1*) gene family are highly induced in response to a wide range of pathogens. Homologues have been identified in barley, but not in *Brachypodium*, whereas in rice, only distant WIR1 candidates are known. Phylogenetic analysis placed TaWIR1a and TaWIR1b within a distinct clade of wheat transcripts, whereas TaWIR1c clustered with *HvWIR1* genes (Tufan et al., 2012). Transcripts of all three *TaWIR1* genes were strongly induced by a wheat-adapted isolate of MoT. Virus-induced gene silencing of the *TaWIR1* gene family had no effect on the initial penetration of epidermal cells by MoT. However, following the establishment of an infection site, MoT was able to grow more extensively within the leaf tissue, relative to control leaves, indicating a role for the *TaWIR1* gene family in the cell-to-cell movement of the fungus. Differential transcription of *TaWIR1* genes was also noticed in epidermal peels, relative to the remaining leaf tissue, during the infection process of MoT. Ha et al. (2016) studied the infection of wheat ears by MoT and *Fusarium graminearum* causing *Fusarium* head blight (FHB). Most of the 27 wheat cultivars inoculated with MoT and *F. graminearum* displayed inverse disease responses to blast and FHB. Two cultivars (Milan and Sumai 3) expressing converse disease phenotypes to blast (Milan and Sumai, resistant and susceptible, respectively) and FHB (Milan and Sumai 3, susceptible and resistant, respectively) were selected. Confocal laser scanning microscopy revealed early (12 hai) colonisation of the spikelets by *F. graminearum* similarly on both cultivars, while anthers of the susceptible cultivar were infected earlier by *F. graminearum*. Both pathogens grew much faster in the rachilla of susceptible than resistant cultivars indicating that resistance is mainly expressed in the part connecting the spikelet with the rachis. Genes for chitinase (Chi2), β-1,3-glucanase (PR2), a plant defensin homologue (PRPI) and peroxidase (Pox2) were strongly up-regulated in Milan in response to both pathogens, while PR2 and PR5 (thaumatin-like protein) were transiently triggered by *F. graminearum* on both cultivars. Up-regulation of cinnamoyl-CoA reductase (CCR), cytochrome P450 (CYP709C1) and UDP-glycosyl transferase (UGT) was more prominent in ears infected with *F. graminearum*, while up-regulation of UGT was higher in Sumai 3 when infected with either pathogen. Cultivar resistance to FHB was reflected by clearly higher expression levels of UGT and CYP709C1 in Sumai 3. The differential responses of wheat to the two ear pathogens make it unlikely that common resistance genes exist for controlling blast and FHB, suggesting the need to stack many genes associated with resistance in breeding programmes for multiple resistance.

REFERENCES

Alves, K. and J. Fernandes. "Influência da temperatura e da umidade relativa do ar na esporulação de *Magnaporthe grisea* em trigo." *Fitopatologia Brasileira* 31 (2006): 579–84.

Araujo, L., J. M. Soares, M. C. C. Filippi and F. A. Rodrigues. "Cytological Aspects of Incompatible and Compatible Interactions Between Rice, Wheat and the Blast Pathogen *Pyricularia oryzae*." *Scientia Agricola* 73 (2016): 177–83.

Boyd, L. A., C. Ridout, D. M. O'Sullivan, J. E. Leach and H. Leung. "Plant–Pathogen Interactions: Disease Resistance in Modern Agriculture." *Trends in Genetics* 29 (2013): 233–40.

Braun, E. J. and R. J. Howard. "Adhesion of Fungal Spores and Germlings to Host Surface." *Protoplasma* 181 (1994): 202–12.

Cardoso, C. A. A., E. M. Reis and E. N. Moreira. "Development of a Warning System for Wheat Blast Caused by *Pyricularia grisea*." *Summa Phytopathologica* 34 (2008): 216–21.

Castroagudín, V. L., S. I. Moreira, D. A. S. Pereira et al. "*Pyricularia graminis-tritici*, the New *Pyricularia* Species Causing Wheat Blast." *Persoonia* 37 (2016): 199–216.

Ceresini, P. C., V. L. Castroagudín, F. A. Rodrigues et al. "Wheat Blast: Past, Present, and Future." *Annual Review of Phytopathology* 56 (2018): 427–56.

Cruz, C. D., J. Kiyuna, W. W. Bockus, T. C. Todd, J. P. Stack and B. Valent. "*Magnaporthe oryzae* Conidia on Basal Wheat Leaves as a Potential Source of Wheat Blast Inoculum." *Plant Pathology* 64 (2015a): 1491–8.

Cruz, C. D., W. W. Bockus, J. P. Stack et al. "Preliminary Assessment of Resistance Among U.S. Wheat Cultivars to the *Triticum* Pathotype of *Magnaporthe oryzae*." *Plant Disease* 96 (2012): 1501–5.

Cruz, M. F. A., J. A. Rios, L. Araujo and F. A. Rodrigues. "Infection Process of *Pyricularia oryzae* on the Leaves of Wheat Seedlings." *Tropical Plant Pathology* 41 (2016): 123–7.

Cruz, M. F. A., L. A. F. Silva, J. A. Rios, D. Debona and F. A. Rodrigues. "Microscopic Aspects of the Colonization of *Pyricularia oryzae* on the Rachis of Wheat Plants Supplied with Silicon." *Bragantia* 74 (2015b): 207–14.

Dean, R. A. "Signal Pathways and Appressorium Morphogenesis." *Annual Review of Phytopathology* 35 (1997): 211–34.

Debona, D., F. A. Rodrigues, J. A. Rios and K. J. T. Nascimento. "Biochemical Changes in the Leaves of Wheat Plants Infected with *Pyricularia oryzae*." *Phytopathology* 102 (2012): 1121–9.

Ebbole, D. J. "*Magnaporthe* as a Model for Understanding Host–Pathogen Interactions." *Annual Review of Phytopathology* 45 (2007): 437–56.

Gomes, D. P., V. S. Rocha, O. L. Pereira and M. A. Souza. "Damage of Wheat Blast on the Productivity and Quality of Seeds as a Function of the Initial Inoculum in the Field." *Journal of Seed Science* 39 (2017): 66–74.

Goulart, A. and F. A. Paiva. "Sobrevivência de *Pyricularia oryzae* Cav. em sementes de trigo (*Triticum aestivum* L.) armazenadas em diferentes ambientes." *Revista Brasileira de Sementes* 15 (1993): 153–6.

Goulart, A., F. Paiva and P. Andrade. "Relação entre a incidência da brusone em espigas de trigo e a presença de *Pyricularia grisea* nas sementes colhidas." *Fitopatologia Brasileira* 20 (1995): 184–9.

Goulart, A. C. P., P. G. Sousa and A. S. Urashima. "Danos em trigo causados pela infecção de *Pyricularia grisea*." *Summa Phytopathologica* 33 (2007): 358–63.

Ha, X., B. Koopmann and A. von Tiedemann. "Wheat Blast and Fusarium Head Blight Display Contrasting Interaction Patterns on Ears of Wheat Genotypes Differing in Resistance." *Phytopathology* 106 (2016): 270–81.

Howard, R. J. "Cell Biology of Pathogenesis." In *Rice Blast Disease*, edited by R. S. Zeigler, S. A. Leong and P. S. Teng, 3–22. Wallingford: Research Institute, CAB International, 1994.
Howard, R. J. and B. Valent. "Breaking and Entering: Host Penetration by the Fungal Rice Blast Pathogen *Magnaporthe grisea*." *Annual Review of Microbiology* 50 (1996): 491–512.
Igarashi, S., C. Utimada, L. Igarashi, A. Kazuma and R. Lopes. "*Pyricularia* sp. em trigo. I. Ocorrência de *Pyricularia* sp. no estado do Paraná." *Fitopatologia Brasileira* 11 (1986): 351–2.
Inoue, K., H. Kitaoka, P. Park and K. Ikeda. "Novel Aspects of Hydrophobins in Wheat Isolate of *Magnaporthe oryzae*: Mpg1, but not Mhp1, is Essential for Adhesion and Pathogenicity." *Journal of General Plant Pathology* 82 (2016): 18–28.
Inoue, K., T. Onoe, P. Park and K. Ikeda. "Enzymatic Detachment of Spore Germlings in *Magnaporthe oryzae*." *FEMS Microbiology Letters* 323 (2011): 13–19.
Inoue, K., T. Suzuki, K. Ikeda et al. "Extracellular Matrix of *Magnaporthe oryzae* May Have a Role in Host Adhesion During Fungal Penetration and is Digested by Matrix Metalloproteinases." *Journal of General Plant Pathology* 73 (2007): 388–98.
Islam, M. T., D. Croll, P. Gladieux et al. "Emergence of Wheat Blast in Bangladesh was Caused by a South American Lineage of *Magnaporthe oryzae*." *BMC Biology* 14 (2016): 11.
Kohli, M. M., Y. R. Mehta, E. Guzman, L. De Viedma and L. E. Cubilla. "*Pyricularia* Blast: A Threat to Wheat Cultivation." *Czech Journal of Genetics and Plant Breeding* 47 (2011): 130–4.
Mendgen, K. and H. Deising. "Infection Structures of Fungal Plant Pathogens: A Cytological and Physiological Evaluation." *New Phytologist* 124 (1993): 193–213.
Mendgen, K., M. Hahn and H. Deising. "Morphogenesis and Mechanisms of Penetration by Plant Pathogenic Fungi." *Annual Review of Phytopathology* 34 (1996): 367–86.
Prabhu, A. S., M. C. Filippi and N. Castro. "Pathogenic Variation Among Isolates of *Pyricularia oryzae* Infecting Rice, Wheat and Grasses in Brazil." *Tropical Pest Management* 38 (1992): 367–71.
Reis, E. M., M. M. C. Blum and C. A. Forcelini. "Sobrevivência de *Pyricularia oryzae* associada a sementes de trigo." *Summa Phytopathologica* 21 (1995): 43–4.
Ribot, C., J. Hirsch, S. Balzergue et al. "Susceptibility of Rice to the Blast Fungus, *Magnoporthe grisea*." *Journal of Plant Physiology* 165 (2008): 114–24.
Rios, J. A., D. Debona, H. S. S. Duarte and F. A. Rodrigues. "Development and Validation of a Standard Area Diagram Set to Assess Blast Severity on Wheat Leaves." *European Journal of Plant Pathology* 136 (2013): 603–11.
Rios, J. A., V. S. Rios, P. A. Paul, M. A. Souza, L. Araujo and F. A. Rodrigues. "Fungicide and Cultivar Effects on the Development and Temporal Progress of Wheat Blast Under Field Conditions." *Crop Protection* 89 (2016): 152–60.
Rios, J. A., V. S. Rios, P. A. Paul, M. A. Souza, L. B. M. C. Neto and F. A. Rodrigues. "Effects of Blast on Components of Wheat Physiology and Grain Yield as Influenced by Fungicide Treatment and Host Resistance." *Plant Pathology* 66 (2017): 877–89.
Rodrigues, F. A., N. Benhamou, L. E. Datnoff, J. B. Jones and R. R. Bélanger. "Ultrastructural and Cytochemical Aspects of Silicon Mediated Rice Blast Resistance." *Phytopathology* 93 (2003): 535–46.
Rodrigues, F. A., W. M. Jurick, L. E. Datnoff, J. B. Jones and J. A. Rollins. "Silicon Influences Cytological and Molecular Events in Compatible and Incompatible Rice–*Magnaporthe grisea* Interactions." *Physiological and Molecular Plant Pathology* 66 (2005): 144–59.
Rodrigues, F. A., J. A. Rios, D. Debona and C. E. Aucique-Pérez. "*Pyricularia oryzae*–Wheat Interaction: Physiological Changes and Disease Management Using Mineral Nutrition and Fungicides." *Tropical Plant Pathology* 42 (2017): 223–9.

Toledo, J. "Enfermedades." In *Guía de Recomendaciones Técnicas del Cultivo de Trigo*, edited by O. Díaz, E. Guzmán, J. Toledo, P. Franco, D. Terrazas and A. Antezana. Santa Cruz, CA: Bolivia, 2004.

Tucker, S. L. and N. J. Talbot. "Surface Attachment and Pre-Penetration Stage Development by Plant Pathogenic Fungi." *Annual Review of Phytopathology* 39 (2001): 385–417.

Tufan, H. A., G. R. D. Mcgrann, R. Maccormack and L. A. Boyd. "TaWIR1 Contributes to Post-Penetration Resistance to *Magnaporthe oryzae*, But Not *Blumeria graminis* f. sp. *tritici*, in Wheat." *Molecular Plant Pathology* 13 (2012): 653–65.

Tufan, H. A., G. R. D. McGrann, A. Magusin, J. B. Morel, L. Miché and L. A. Boyd. "Wheat Blast: Histopathology and Transcriptome Reprogramming in Response to Adapted and Nonadapted *Magnaporthe* Isolates." *New Phytologist* 184 (2009): 473–84.

Urashima, A. S., S. F. Leite and R. Galbieri. "Eficiência da disseminação aérea em *Pyricularia grisea*." *Summa Phytopathologica* 33 (2007): 275–9.

Urashima, A. S., T. D. Martins, C. R. N. C. Bueno, D. B. Favaro, M. A. Arruda and Y. R. Mehta. "Triticale and Barley: New Hosts of *Magnaporthe grisea* in São Paulo, Brazil – Relationship with Blast of Rice and Wheat." In *Rice Blast: Interaction with Rice and Control*, edited by S. Kawasaki, 251–60. Dordrecht: Kluwer Academic, 2004.

Urashima, A. S., C. R. F. Grosso, A. Stabili et al. "Effect of *Magnaporthe grisea* on Seed Germination, Yield and Quality of Wheat." In *Advances in Genetics, Genomics and Control of Rice Blast Disease*, edited by G. L. Wang and B. Valent, 267–77. Dordrecht: Springer, 2009.

Valent, B. and F. G. Chumley. "Molecular Genetic Analysis of the Rice Blast Fungus, *Magnaporthe grisea*." *Annual Review of Phytopathology* 29 (1991): 443–67.

Van Vu, B., K. Itoh, Q. B. Nguyen, Y. Tosa and H. Nakayashiki. "Cellulases Belonging to Glycoside Hydrolase Families 6 and 7 Contribute to the Virulence of *Magnaporthe oryzae*." *Molecular Plant-Microbe Interactions* 25 (2012): 1135–41.

Veneault-Fourrey, C. and N. J. Talbot. "Moving Toward a Systems Biology Approach to the Study of Fungal Pathogenesis in the Rice Blast Fungus *Magnaporthe grisea*." *Advances in Applied Microbiology* 57 (2005): 177–215.

Veneault-Fourrey, C., M. Barooah, M. Egan, G. Wakley and N. J. Talbot. "Autophagic Fungal Cell Death is Necessary for Infection by the Rice Blast Fungus." *Science* 312 (2006): 580–3.

Wang, X., F. X. Shi, H. A. B. Wösten, H. Hektor, B. Poolman and G. T. Robillard. "The SC3 Hydrophobin Self-Assembles Into a Membrane with Distinct Mass Transfer Properties." *Biophysical Journal* 88 (2005): 3434–43.

Wilson, R. A. and N. J. Talbot. "Under Pressure: Investigating the Biology of Plant Infection by *Magnaporthe oryzae*." *Nature Reviews Microbiology* 7 (2009): 185–95.

Wösten, H. A. B. "Hydrophobins: Multipurpose Proteins." *Annual Review of Microbiology* 55 (2001): 625–46.

Xiao, J. Z., A. Ohshima, T. Kamakura, T. Ishiyama and I. Yamaguchi. "Extracellular Glycoprotein(s) Associated with Cellular Differentiation in *Magnaporthe grisea*." *Molecular Plant-Microbe Interactions* 7 (1994): 639–44.

5 Genetic and Molecular Basis of Wheat–*Magnaporthe oryzae Triticum* Interaction

Jéssica Rosset Ferreira, Gisele Abigail Montan Torres, Luciano Consoli, Eliseu Binneck, Gabriela Andriolio Camilotti, Sandra Maria Mansur Scagliusi, Carolina Cardoso Deuner, Alexei de Campos Dianese, Augusto César Pereira Goulart, Claudine Dinali Santos Seixas and Maurício Antônio de Oliveira Coelho

CONTENTS

5.1 INTRODUCTION

The first world report of wheat blast occurrence in field conditions was in northern Parana, Brazil, in the mid-1980s (Igarashi et al., 1986). In subsequent years, the pathogen spread to new areas, and its presence has been registered in the Distrito Federal (Anjos et al., 1996) and in the states of Mato Grosso do Sul (Goulart et al., 1990), Rio Grande do Sul (Picinini & Fernandes, 1990), São Paulo (Igarashi, 1991),

Goiás (Prabhu et al., 1992) and Minas Gerais. Blast prevails in all the wheat-producing regions of Brazil due to the favourable conditions of high temperatures (above 20°C) and humidity (over 10 hours of leaf wetting) (Cardoso et al., 2008).

In addition to Brazil, the disease was observed on the Brazil/Paraguay border in 1987, and reached epidemic proportions in Paraguay in 2002 (Viedma & Paiva, 2002). In the lowlands of Santa Cruz, Bolivia, severe blast infections were observed in 1996, causing approximately 80% yield losses (Barea & Toledo, 1996). Subsequently it was first identified in Argentina in 2007, limited to the north of the country (Cabrera & Gutiérrez, 2007). In Uruguay, infections have been sporadic especially during the wet and warm years (Kohli et al., 2011). In 2011, wheat blast was also found in North America. A single spike of wheat with blast symptoms was observed in Princeton, Kentucky (Farman et al., 2017). In February 2016, blast was recorded in southern Bangladesh in Asia, reaching approximately 15% of the area cultivated with wheat (Malaker et al., 2016).

Blast disease or *brusone* is caused by the ascomycete *M. oryzae* B.C. Couch and L.M. Kohn (synonym *Pyricularia oryzae* Cavara) which infects more than 100 species of the Poaceae family including rice, wheat and barley (Choi et al., 2013). *M. oryzae* can infect all the aerial parts of the wheat plant, including the leaves, stems and spikes. Spike infection is the most destructive form of the disease (Prestes et al., 2007). The damage caused by blast to wheat grain production depends on the time of spike infection and the site of penetration of the fungus; it can affect the spike partially or completely.

The importance of blast disease arises from the reductions in yield and grain quality. In early infections, newly formed wheat grains are deformed, small and with low specific weight. Most of them are eliminated in the harvesting and processing. In the state of Mato Grosso do Sul, research carried out from 1988 to 1992, under natural conditions of infection in the field, showed a reduction of 10%–53% in grain yield and 14.5%–74% in the weight of the grains produced (Goulart et al., 2007). Diseased spikes have characteristic discolouration or bleaching, and the rachises exhibit a necrosis at the site of penetration of the fungus. Spikes bleaching symptoms are so impressive that damages due to blast disease could be overestimated (Goulart et al., 2007).

5.2 GENETIC RESISTANCE

Wheat blast caused by *M. oryzae Triticum* was first reported in 1985 in Parana, Brazil, whereas rice blast was described in 1673 in China (Chiu & Chang, 1982) and is the most studied host pathosystem. Rice blast has become a model system for molecular genetic studies about plant–pathogen interaction (Liu et al., 2010a), and can help in understanding the wheat *M. oryzae Triticum* pathosystem. Under this temporal perspective, there is a clear lack of knowledge about the interaction of *M. oryzae* with wheat in relation to its interaction with rice.

Blast resistance in rice has been classified as qualitative/complete or quantitative/partial resistance. Complete resistance is controlled by single genes that confer specific resistance on a particular or a few pathogen races of *M. oryzae* (R genes). Quantitative or partial resistance is conferred by quantitative trait loci (QTL) or is associated with several genes with major and minor effects. Over 100 blast resistance genes have been identified and 25 have been cloned and characterised (Ashkani et al., 2016;

Devanna & Sharma, 2018). For rice blast, it was demonstrated that race-cultivar specificity follows the gene-for-gene relationship (Silué et al., 1992). For each gene in plant that confers resistance to the pathogen there is a corresponding gene in the pathogen that confers pathogenicity to the host (Flor, 1956).

Besides the cultivar specificity, gene-for-gene interactions underlie some specificities between plant species and fungal pathogens, which is considered as host species specificity (Tosa, 2009). In wheat, knowledge about genetic resistance comes mainly from the crossing of fungal strains belonging to different host species. A research group from Kobe University, Japan, performed a "pivot turn crossing" of subgroups of *Magnaporthe*. Following this approach, pathogen avirulence genes with the corresponding putative resistance genes in wheat could be identified.

To date, nine resistance genes to blast have been identified in wheat (Table 5.1). Four are effective against isolates obtained from other grasses. *Rmg1* (synonymous with *Rwt4*) is the first-known wheat gene for resistance to *M. oryzae* identified in Norin 4 inoculated at the seedling stage. It confers resistance to *Avena* (Takabayashi et al., 2002) and *Panicum* isolates (Hau et al., 2007), but is ineffective against *Triticum* isolates. In *Avena* isolate Br58, two avirulence genes named *PWT3* and *PWT4* condition the avirulence on wheat. *Rmg1* (*Rwt4*) recognises *PWT4* (Takabayashi et al., 2002) under the control of the gene-for-gene relationship.

PWT3 and *PWT4* were isolated and the predicted proteins contained putative signal peptides but with no similarity to known proteins or protein domains (Inoue et al., 2017). Both genes were used to transform *Triticum* isolate Br48, which was virulent on the spikes of a set of wheat cultivars. The transformants with *PWT3* or *PWT4* were specifically avirulent on the cultivars carrying *Rmg6* or *Rmg1*, respectively. The interactions *Rmg6(Rwt3)–PWT3* and *Rmg1(Rwt4)–PWT4* play roles in seedlings and spike infection on wheat.

Rmg4 and *Rmg5* conferred resistance to a *Digitaria* isolate in Norin 4 and Red Egyptian, respectively (Nga et al., 2009). *Rmg4* and *Rmg5* are located on chromosomes 4A and 6D, respectively. *Rmg6* was identified in Norin 4 and is effective against a *Lolium* isolate but not against *Triticum* isolates (Vy et al., 2014). The avirulence of *Lolium* isolate TP2 on wheat cultivars Chinese Spring, Norin 4 and Shinchunaga is conditioned by two unlinked genes, designated *A1* and *A2*. Both genes are epistatic. *A1* is involved in strong avirulence and *A2* is involved in moderate avirulence. The *A1* is located at the same locus as *PWT3* and is recognised by *Rmg6* (Inoue et al., 2017). So, the incompatibility of the *Lolium* isolate and wheat is conditioned by *Rmg6–PWT3* and *R2–A2* gene pairs in a gene-for-gene manner as it occurs for *Avena* isolates.

Rmg2 and *Rmg3* were identified in cultivar Thatcher against *Triticum* isolates. Both are effective at the seedling stage and are located on chromosomes 7A and 6B, respectively (Zhan et al., 2008). *Rmg7* was identified in tetraploid accessions St 24, St 17 and St 25 and conferred resistance at the seedling as well as the heading stage against a *Triticum* isolate. *Rmg7* is located on the long arm of chromosome 2A and its corresponding avirulence gene is designated *AVR-Rmg7* (Tagle et al., 2015). Finally, *Rmg8* was identified on hexaploid wheat S615, located on the long arm of chromosome 2B. The resistance conferred by *Rmg8* is expressed at the seedling and heading stages. The corresponding avirulence gene was identified as *AVR-Rmg8* (Tagle et al., 2015).

TABLE 5.1
Summary of Wheat Resistance Genes to *Magnaporthe* sp. Isolates

Resistance gene	Host plant species	Accession	Chr	Expression of resistance	Avirulence gene	*Magnaporthe* sp. isolate	References
Rmg1 (*Rwt4*)	*T. aestivum* (hexaploid)	Norin 4	1D	Seedling and heading stage	*PWT3*, *PWT4*	From *Avena*	Takabayashi et al. (2002); Hirata et al. (2005); Inoue et al. (2017)
Rmg2	*T. aestivum* (hexaploid)	Thatcher	7A	Seedling stage, not at heading stage	–	From wheat	Zhan et al. (2008)
Rmg3	*T. aestivum* (hexaploid)	Thatcher	6B	Seedling stage, not at heading stage	–	From wheat	Zhan et al. (2008)
Rmg4	*T. aestivum* (hexaploid)	Norin 4	4A	Seedling stage	–	From *Digitaria*	Nga et al. (2009)
Rmg5	*T. aestivum* (hexaploid)	Red Egyptian	6D	Seedling stage	–	From *Digitaria*	Nga et al. (2009)
Rmg6 (*Rwt3*)	*T. aestivum* (hexaploid)	Chinese Spring, Norin 4, Shin-chunaga	1DS	Seedling and heading stage	*PWT3*	From *Lolium*	Vy et al. (2014); Inoue et al. (2017)
Rmg7	*T. dicoccum* (tetraploid)	St 24 (KU120), St 17 (KU112), St 25 (KU122)	2AL	Seedling and heading stage	*AVR-Rmg7* synonym *AVR-Rmg8*	From wheat	Tagle et al. (2015); Anh et al. (2017)
Rmg8	*T. aestivum* (hexaploid)	S615	2BL	Seedling and heading stage	*AVR-Rmg8*	From wheat	Anh et al. (2017); Anh et al. (2015)
RmgGR119	*T. aestivum* (hexaploid)	GR119	–	Seedling and heading stage	*AVR-Rmg8*, *AVR-RmgGR119*	From wheat	Wang et al. (2018)
RmgTd(t)	*T. dicoccoides* (tetraploid)	Tat4	–	Against white cultures of isolates	–	Backcrossing of an *Avena* and a *Triticum* isolate	Cumagun et al. (2014)

Recently, *AVR-Rmg7* and *AVR-Rmg8* have been shown to be the same gene recognised by two resistance genes, *Rmg7* in tetraploid wheat and *Rmg8* in hexaploid wheat (Anh et al., 2017). *Rmg7* and *Rmg8* differ in their chromosomal location and in their temperature sensitivity. At room temperature (21°C–24°C), both genes are effective against *Triticum* isolates. However, at 26°C, *Rmg7* is ineffective whereas *Rmg8* is still effective. The relationship between *AVR-Rmg8* and *Rmg7*/*Rmg8* appears to be a gene-for-genes interaction (DeFeyter et al., 1993), in which one avirulence gene corresponds to more than one resistance gene.

Among 520 landraces of wheat collected worldwide, one highly resistant accession GR119 was identified. This genotype carries two resistance genes effective against wild isolates of *M. oryzae* obtained from wheat collected in Brazil. Both genes act additively, *Rmg8* and a new gene proposed to be named *RmgGR119*, which is strongly resistant in seedlings and at heading stage (Wang et al., 2018).

In the tetraploid wheat (*Triticum dicoccoides*) Tat4, a single major gene was found that conferred a moderate level of resistance (Cumagun et al., 2014). It was designated as *RmgTd*(t) and it is considered a hidden resistance gene. It confers resistance against mutant white cultures of *M. oryzae*, but it does not confer resistance against field isolates. The gene responsible for the pigment and virulence defects seems to be an avirulence gene corresponding to *RmgTd*(t) in a gene-for-gene interaction.

Besides specific genes conferring resistance to specific isolates of *Magnaporthe* sp., an association between wheat resistance to blast and the presence of a chromosomal segment (2NS) from *Triticum ventricosum* (syn. *Aegilops ventricosa*) was found (Cruz et al., 2016). A total of 418 wheat cultivars were characterised in inoculation with *M. oryzae*, including winter, spring and quasi-isogenic lines. Some of the materials were also tested under field conditions in Bolivia. In both types of experiments, there were lower incidences of blast in accessions with the 2NS/2AS translocation. Not all wheat genotypes carrying the 2NS/2AS translocation showed a significant reduction in spike blast severity. This suggests that the genetic background and/or environment may influence the expression of the resistance conferred by this translocation (Cruz et al., 2016).

In Brazil, since the appearance of blast, many studies have been carried out aimed at identifying the resistance sources to the pathogen. Due to pathogen variability, it is essential to characterise wheat cultivars for different pathogen populations. The Brazilian Wheat and Triticale Research Commission publishes annual data about the wheat cultivars indicated for cultivation in the country (Silva et al., 2017). Between 2005 and 2017, 228 cultivars were qualitatively evaluated for different diseases under Brazilian regions of cultivation. Most of the cultivars (83.8%) remained under cultivation for more than one consecutive year. However, the threat imposed on wheat production persists because for 48.2% of the cultivars there is no information about reaction to the blast disease (Table 5.2). Only two cultivars (BRS 229 and Trigo BR18-Terena) are described as resistant to the disease.

BRS 229 (line WT 96168) is a cultivar from Embrapa Soja that was under cultivation between 2004 and 2012 and had good resistance to both the main fungal diseases and to the blast disease. Trigo BR18-Terena, whose pedigree is unknown, was originally introduced from CIMMYT by Embrapa Trigo. "BR 18" was released in 1986, one year after the emergence of wheat blast in Brazil. Actually, it remains

TABLE 5.2
Types of Reaction to Blast Disease and Number of Wheat Cultivars Indicated for Cultivation in Brazil, According to the Brazilian Wheat and Triticale Research Commission, from 2005 to 2017

Types of reaction*	Number of cultivars	%
Highly susceptible (HS)	1	0.4
Susceptible (S)	27	11.8
Moderately susceptible/susceptible (MS/S)	1	0.4
Moderately susceptible (MS)	32	14.0
Moderately susceptible/moderately resistant (MS/MR)	5	2.2
Moderately resistant/moderately susceptible (MR/MS)	4	1.8
Moderately resistant (MR)	45	19.7
Moderately resistant/resistant (MR/R)	1	0.4
Resistant (R)	2	0.9
No information	110	48.2
Total	268	

*The most common reaction for wheat cultivars indicated for consecutive years of cultivation.

indicated for cultivation in the country especially due to its blast resistance and bread-making quality performance.

A collection of wide variability of relevant accessions was composed for the purpose of breeding programmes in Brazil. A total of 196 genotypes were evaluated in wheat blast nurseries under natural infections. Three sites, Planaltina-DF, Dourados-MS and Londrina-PR, and three years of evaluation were considered for analysis. Fifteen wheat genotypes presented lower incidences of disease and were selected for further studies having as susceptibility controls Anahuac 75 and BRS 209 (Table 5.3). Only two of them, CBFusarium ENT014 and CPAC 07434, possessed the 2NS/2AS translocation (Ferreira et al., 2018). Among 13 genotypes without the 2NS/2AS translocation, Trigo BR 18-Terena and BRS 229 are two of the most important wheat cultivars in Brazil with durable resistance to blast.

Moreover, a restricted set of wheat genotypes, all with the 2NS/2AS translocation, showed a wide variability of reactions to the disease (Pizolotto et al., 2017). These data indicated that the sole presence of this translocation does not necessarily imply spike resistance. Further studies will be necessary to evaluate the contribution of this translocation to the resistance reaction observed in the two genotypes, CBFusarium ENT014 and CPAC 07434, under the conditions of wheat cultivation in Brazil. The identification of resistant materials that do not possess the 2NS/2AS translocation (such as BRS 229 and BR 18) opens up prospects for the study and discovery of genes and/or QTL and mechanisms of resistance not related to this translocation and that are effective for the Brazilian conditions where blast is endemic.

Pathogen can infect all wheat aerial tissues, producing lesions of different shapes and sizes. Rice seedlings reaction to *M. oryzae* is usually classified as susceptible or

TABLE 5.3
Wheat Genotypes Selected for Analysis of the Presence of 2NS/2AS Translocation (Embrapa Trigo, Passo Fundo, RS, Brazil, 2017)

Wheat genotype	Germplasm bank code	Pedigree	Release date	Country	Blast reaction in hotspots	2NS/2AS translocation
Anahuac 75	BGT00443	II-12300//Lerma-Rojo-64/II-8156/3/Norteno-67	1975	Mexico	Susceptible	–
Bet Dagan 131	BGT01227	Pitic-62(SIB)/Florence-Aurore	1970	Iceland	Resistant	–
BRS 209	BGT14405	Jupateco 73/Embrapa 16	2002	Brazil	Susceptible	–
BRS 229	BGT14408	Embrapa 27*3//BR 35/Buck Poncho	2004	Brazil	Resistant	–
BRS Angico	BGT01454	PF 87107/2*IAC 13	2002	Brazil	Resistant	–
CBFusarium ENT014	BGT15263	No information	–	Mexico	Resistant	+
CPAC 07340	BGT20389	CPAC 96306/CPAC 9985	–	Brazil	Resistant	–
CPAC 07434	BGT20388	Taurum/BRS 254	–	Brazil	Resistant	+
Embrapa 27	BGT03372	PF 83743/5/PF 83182/4/CNT 10*4//Lagoa Vermelha*5/ Agatha/3/Londrina*4/Agent//Londrina*3/Nyu Bay	1994	Brazil	Resistant	–
Huanca	BGT04275	Frocor/3/McMurachy/Kentana//Yaqui-50/4/Maria-Escobar/ MN-2698/5/Maria-Escobar	1973	Peru	Resistant	–
PF 020450	BGT16243	FL 72185A-A2-C1/Embrapa 40//CEP 24	–	Brazil	Resistant	–
PF 909	BGT11109	PF 83743/PF 82252//PF 84433/BR 35	–	Brazil	Resistant	–
Safira	BGT15079	PF9099 /OR-1//Granito	2003	Brazil	Resistant	–
Shanghai	BGT15211	(M)Yangmai-1	1978	China	Resistant	–
Thatcher	BGT13324	Marquis/Iumillo(durum)//Marquis/Kanred	1934	USA	Resistant	–
Trigo BR 18-Terena	BGT13539	No information	1986	Brazil	Resistant	–
Trigo Chapéu	BGT13588	No information	–	–	Resistant	–

– absence; + presence.

FIGURE 5.1 Blast disease symptoms on the rachis. Grains are affected in the upper portion of the spike above the infection point whereas the portion below remains healthy and produces normal grains. Black arrows indicate the infection points on the rachis selected to separate the grains. White arrows indicate several infection points on the rachis.

resistant according to lesion types and the presence of sporulation (Bonman et al., 1986; Valent et al., 1991). In a similar way, there are five classes of infection for the classification of wheat reaction to different isolates (Urashima et al., 1993). However, the mesothetic reaction was commonly observed in adult wheat plants evaluated under controlled conditions (Torres et al., 2016). Even on spikes, before bleaching symptoms, different infection types can be seen on diseased spikelets.

In Brazil, wheat blast is endemic and most of the available cultivars are susceptible to the pathogen. Depending on the wheat cultivar, severe leaf infection occurs under field conditions in Brazil. However, little is known about the role of conidia production from leaves in wheat spike blast development, which has the most drastic effect on grain production. Spike bleaching is the typical symptom of blast and is easily confounded with *Fusarium* head blight (causal agent *Fusarium* spp.). When *M. oryzae* infects the wheat rachis, it prevents sap flow to grain filling. Above the penetration site of the pathogen, commonly named the "infection point", grains are affected, whereas the portion below remains healthy and produces normal grains (Figure 5.1 – black arrows).

An important feature is that some wheat genotypes under high pressure of disease may present several infection points on the rachis (ipr) (Toledo, 2004). Upon inoculation, it is common to find variation in the number of ipr. This pattern has already been observed under field conditions (Figure 5.1 – white arrows). Nevertheless, as far as we know, the effect of a few or many ipr on grain production has never been analysed.

5.3 BRS 209 × CBFUSARIUM ENT014 PHENOTYPES OF REACTION TO BLAST: A CASE STUDY

Even though it is easier to test seedling reaction to *M. oryzae* infection, there is no correspondence between the resistance of young and adult plants or between flag leaves and spikes (Maciel et al., 2014). For wheat grain production, it is essential to search for and deploy genotypes with a good level of resistance in spikes. Taking this into account, a Brazilian research initiative was undertaken to identify and molecularly characterise wheat genotypes with contrasting blast reaction phenotypes.

A collection of 80 wheat genotypes were inoculated at the heading stage with the *M. oryzae Triticum* isolate Py6025 under controlled conditions. Anahuac 75 was employed as control of a susceptible reaction (Figure 5.2A and B). The cultivar BRS 209, from Embrapa Trigo, was previously described as susceptible to *M. oryzae* under inoculation conditions (Prestes et al., 2007). Its reaction to Py6025 (Figure 5.2D) was similar to the Anahuac 75 susceptibility response (Figure 5.2B). Among the inoculated genotypes, CBFusarium ENT014 was identified as resistant to the pathogen (Figure 5.2E and F). It is a derivative synthetic wheat belonging to a crossing block (CB) from CIMMYT (Van Ginkel et al., 2002) that was introduced into the Germplasm Bank of Embrapa Wheat in 2008. The CB, at CIMMYT, are collections of elite breeding parental stock and source materials arranged by mega-environment (ME). Genotypes from each section of the CB carry specific genes to their defined ME. In this case, the ME is *Fusarium* head blight. As far as we know, CBFusarium ENT014 had not been characterised for blast resistance before. In plants of this genotype, disease lesions of different types are visible four days after inoculation. However, bleaching develops more slowly compared to that in the susceptible wheat variety BRS 209.

All the inoculated spikes of BRS 209 and of CBFusarium ENT014 presented symptoms of successful infection (Figure 5.2D and F), in contrast to mock-inoculated spikes (Figure 5.2C and E). Blast intensity was measured according to severity quantitation over time since the appearance of the first symptoms. Severity was expressed as the percentage of spikelets with lesions at 7 dpi and as the percentage of bleaching at 21 dpi (Table 5.4).

Blast disease was also expressed as severity of bleaching over time (Figure 5.3). CBFusarium ENT014 presented delayed development of spike bleaching under our

FIGURE 5.2 Macroscopic views of mock-inoculated and inoculated spikes from wheat genotypes Anahuac 75, BRS 209 and CBFusarium ENT014 at 10 days post-inoculation. (A) Mock-inoculated spike from Anahuac 75. (B) Anahuac 75 inoculated with *Magnaporthe oryzae*; the entire spike is bleached. (C) Mock-inoculated spike from BRS 209. (D) BRS 209 inoculated with *M. oryzae*; the entire spike is bleached. (E) Mock-inoculated spike from CBFusarium ENT014 genotype. (F) CBFusarium ENT014 inoculated with *M. oryzae*; showing several spikelets with blast lesions but no bleaching.

TABLE 5.4
Severity (%) of Blast in Wheat Spikes Inoculated with *Magnaporthe oryzae* and Evaluated at 7 and 21 Days Post-Inoculation (dpi)

Wheat genotype	Spikelets with lesions at 7 dpi (%)	Bleached spikelets at 21 dpi (%)
Anahuac 75	48.6	96.4
BRS 209	41.9	99.7
CBFusarium ENT014	48.2	14.2

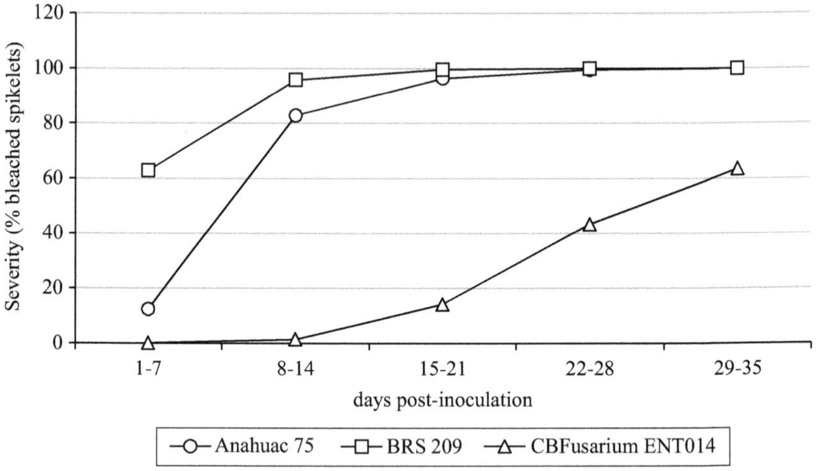

FIGURE 5.3 Progress of severity of blast disease expressed as a percentage of bleached spikelets until 35 days post-inoculation of wheat genotypes Anahuac 75, BRS 209 and CBFusarium ENT014 inoculated with *Magnaporthe oryzae Triticum* isolate Py6025. Seventy-five spikes per genotype were considered.

experimental conditions. Thus, we proposed this genotype as a resistant-candidate that deserves further characterisation.

At maturation, inoculated spikes were harvested and threshed individually. Grain production was evaluated by the number of grains per spike and the average grain weight. For mock-inoculated spikes, only the average grain weight between Anahuac 75 and CBFusarium ENT014 was not statistically different (Figure 5.4). CBFusarium ENT014 presented the highest values for both traits. Inoculated spikes of CBFusarium ENT014 also presented higher values than BRS 209 and Anahuac 75. Comparing the inoculated spikes of susceptible genotypes, BRS 209 had significantly higher number of grains than Anahuac 75 but the average grain weight was similar.

Significant differences between mock-inoculated and inoculated spikes were observed in susceptible genotypes for both traits (Figure 5.4). Mock-inoculated and

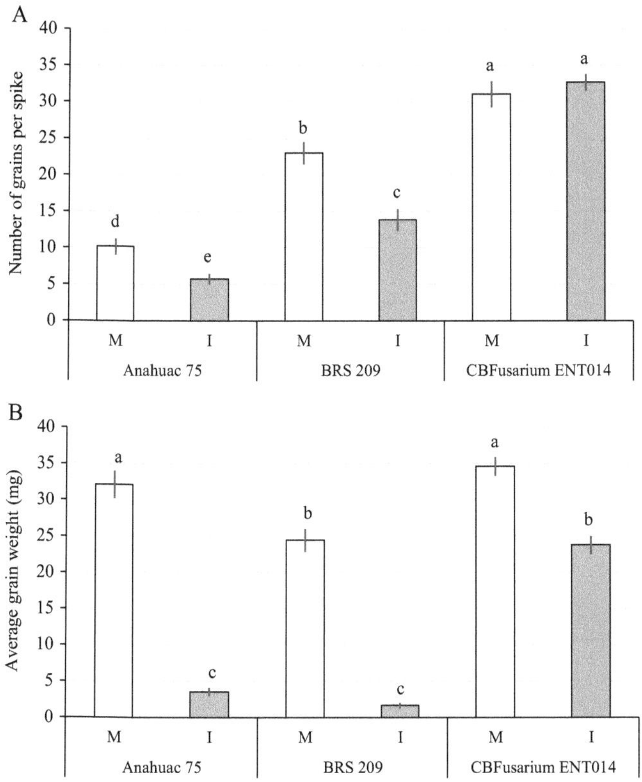

FIGURE 5.4 Number of grains per spike (A), and average grain weight in milligrams (B) of Anahuac 75, BRS 209 and CBFusarium ENT014 inoculated with *Magnaporthe oryzae Triticum* isolate Py6025. The histograms represent the mean ± SE of mock-inoculated (M) spikes (white) and inoculated (I) spikes (grey). On average, 45 spikes per treatment per genotype were considered. Different letters indicate significant differences ($P < 0.05$).

inoculated spikes of CBFusarium ENT014 produced similar number of grains per spike (Figure 5.4A), but the average grain weight was statistically different between both treatments (Figure 5.4B). In this resistant genotype, blast affects grain production by reducing weight but not by reducing the number of grains.

After threshing, the number of ipr of each inoculated spike was counted. The three genotypes were statistically different (Figure 5.5). Interestingly, the two susceptible genotypes showed lower numbers of ipr; BRS 209 had the lowest number of ipr followed by Anahuac 75. CBFusarium ENT014 had an average of 10 ipr.

These results led us to make inferences about the resistance mechanisms to *M. oryzae Triticum*. CBFusarium ENT014 has a slower development of bleaching which suggests resistance to pathogen colonisation. Additionally, it possesses a larger number of ipr, which could be a hypersensitive response (HR) to pathogen infection. A hypersensitive response is a form of localised programmed cell death with restriction on pathogens to the infection site that often lead to the appearance of macroscopically visible localised tissue necrosis (Künstler et al., 2016).

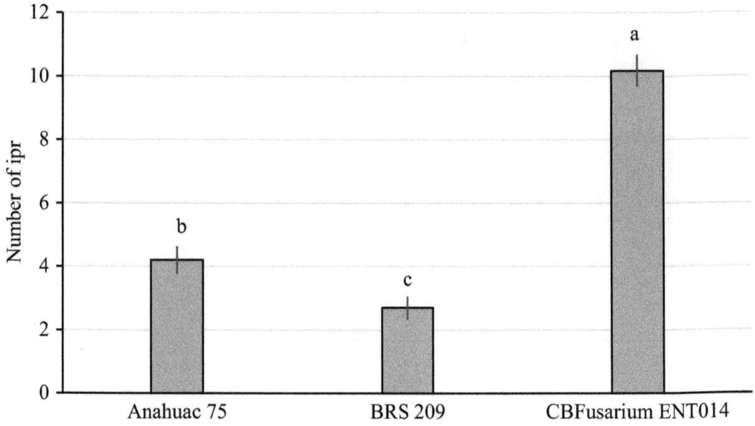

FIGURE 5.5 Average number of infection points on both sides of the rachis identified in wheat genotypes Anahuac 75, BRS 209 and CBFusarium ENT014 inoculated with *Magnaporthe oryzae Triticum* isolate Py6025. The histograms represent the mean ± SE of 40 rachises per genotype on average. Different letters indicate significant differences (P < 0.05).

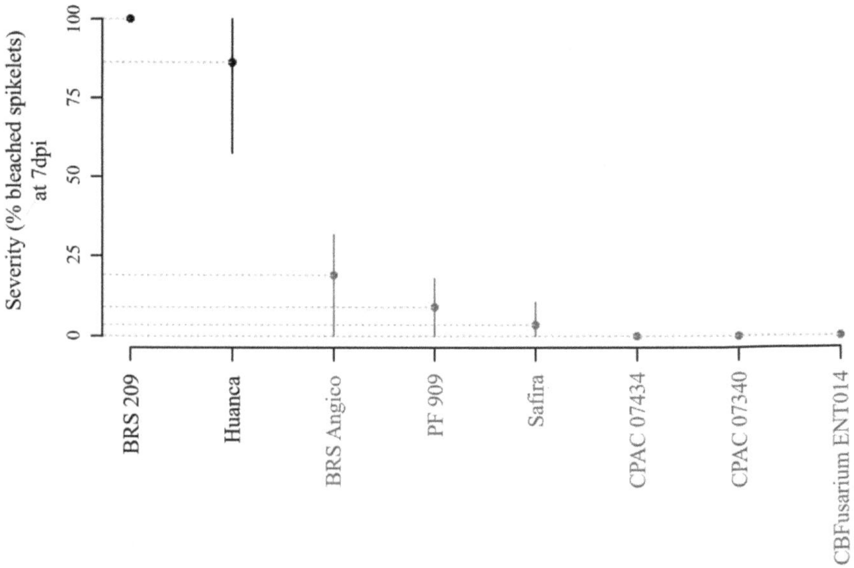

FIGURE 5.6 Severity of blast disease expressed as a percentage of bleached spikelets at 7 days post-inoculation with *Magnaporthe oryzae Triticum* isolate Py0925.

Six other wheat genotypes were inoculated at the heading stage with a different *M. oryzae Triticum* isolate Py0925 to investigate the relation between the number of ipr and resistance. At 7 days post-inoculation, two groups of disease severity were identified (Figure 5.6). BRS 209 and Huanca with severity expressed as a percentage of bleached spikelets near 100% and the others had statistically lower severities in

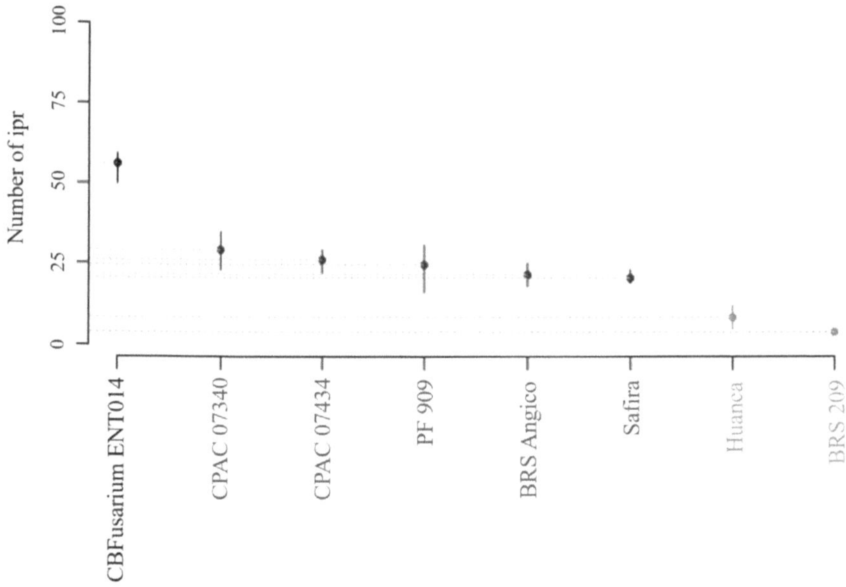

FIGURE 5.7 Total number of infection points on both sides of the rachis identified in wheat genotypes inoculated with *Magnaporthe oryzae Triticum* isolate Py0925.

the range from zero to 20%. After threshing, ipr were counted and variability in the data could be noticed (Figure 5.7). Interestingly, the variables severity of bleaching at 7 dpi and number of ipr presented a negative correlation (r = –0.68). Bleaching was lower for genotypes with more ipr. These data may indicate a coupled resistance mechanism with delayed bleaching and a stronger HR.

BRS 209 and CBFusarium ENT014 are highly contrasting for the traits severity of bleaching and number of ipr. These genotypes were crossed and a doubled-haploid wheat population ("pop Embrapa-brusone") was developed for genetic studies of resistance to blast. Under controlled conditions, parental genotypes, Anahuac 75 and 59 doubled haploid lines were inoculated and analysed for the number of ipr. The average number of ipr varied from 2.9 to 53.1. BRS 209 had an average of 2.9 and CBFusarium ENT014 had 9.3. Anahuac 75 had 4.3 on average. The segregation of this trait across the population could be observed (Figure 5.8). BRS 209 had the lowest average number of ipr on both sides of the rachis and 51 lines (86.4%) presented more ipr than the parent CBFusarium ENT014.

The identification of lines exhibiting reduced bleaching severity with low ipr indicates that these characteristics segregate independently. Genetic mapping of this population is needed to identify QTL for blast disease resistance.

5.4 MOLECULAR PLANT–PATHOGEN INTERACTION

Plants have evolved several mechanisms to defend themselves against diverse pathogens. Plant defence mechanisms include physical barriers, antimicrobial compounds and the innate immune system. Plant innate immunity is cell autonomous, with each

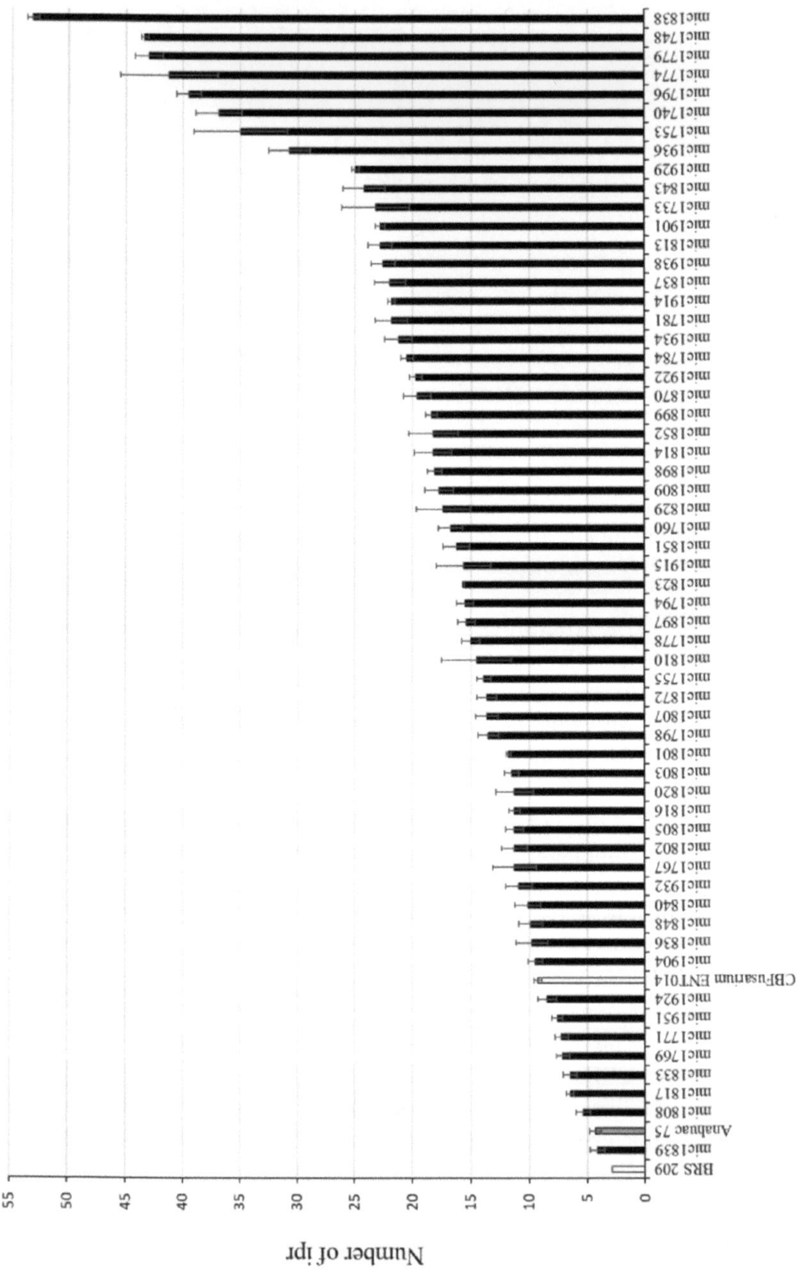

FIGURE 5.8 Average number of infection points on both sides of the rachis identified in doubled-haploid lines derived from the cross between BRS 209 and CBFusarium ENT014 inoculated with *Magnaporthe oryzae Triticum* isolate Py6025 in a 95% confidence interval. Grey bar is the control of a susceptible reaction Anahuac 75. White bars are the parental lines BRS 209 and CBFusarium ENT014. Black bars are the doubled haploid population lines.

cell able to recognise microbial signals and respond accordingly using systemic signals from infection sites (Jones & Dangl, 2006; Cesari et al., 2014). Following pathogen recognition and signal transduction, defence responses that protect plants from infection are activated (Chen & Ronald, 2011). There are two main types of innate immunity in plants: PTI (pathogen-associated molecular patterns, PAMP-triggered immunity) and ETI (effector-triggered immunity).

PTI is governed by the recognition of conserved microbial signatures by host sensors known as PRRs (pattern recognition receptors) (Ronald & Beutler, 2010). These extracellular transmembrane sensors trigger a relatively weak immune response to detect the presence of PAMPs released in the extracellular environment and inhibit colonisation of invading organisms. Plant PRRs also recognise a range of host-derived endogenous signals named danger- or damage-associated molecular patterns (DAMPs) that modulate PTI responses. PRRs are encoded by large gene families and are typically receptor kinases, which include transmembrane receptor-like kinases (RLKs) and transmembrane receptor-like proteins (RLPs) with modular functional domains. RLKs are transmembrane proteins that have amino-terminal ectodomains acting as recognition sites, a transmembrane domain and an intracellular kinase domain. RLPs are RLKs without the intracellular kinase domain. In rice, 1131 RLK family members were identified and are involved in the control of plant growth and development, or in plant–microbe interactions and defence responses (Shiu et al., 2004).

PAMPs are conserved among a large group of pathogens and are considered to have an essential function in microbial survival (Thomma et al., 2011). The PAMPs recognised by plants include proteins, carbohydrates, lipids and small molecules, such as ATP. Most of the known PAMPs can be categorised into polypeptides or carbohydrates (Liu et al., 2013). Polypeptide PAMPs include the bacterial flagellin, elongation factor Tu and Ax21. Carbohydrate PAMPs are components of the microbial cell envelope such as bacterial lipopolysaccharides and peptidoglycan and fungal chitin. In animals, PAMPs and virulence factors are not equivalent (Medzhitov, 2001) because the former evolved to perform physiological functions, whereas virulence factors are produced to interact with the host. In plants, this distinction is not maintained; several PAMPs also have virulence functions (Thomma et al., 2011). Some examples of PAMPs with virulence functions are bacterial flagellin, peptidoglycan, lipopolysaccharides and *M. oryzae* PWL proteins. PWL1 and PWL2 are secreted glycine-rich hydrophilic proteins that prevent infection of weeping lovegrass (Kang et al., 1995; Sweigard et al., 1995).

ETI is governed by the recognition of highly variable microbial molecules secreted into the plant cytoplasm and are known as effectors or avirulence (Avr) effectors. The best-known effectors are from bacteria that deliver effectors directly into host cells. Fungal and oomycete pathogens often secrete effector proteins to the extracellular milieu through the eukaryotic secretory pathway (Panstruga & Dodds, 2009). *M. oryzae* has two secretion systems to deliver effectors in host plants (Giraldo et al., 2013). The first one is apoplastic and accumulates extracellularly at the host–pathogen interface. It is secreted via the process commonly described for fungi. The other involves the exocyst complex and delivers cytoplasmic effectors inside rice cells.

These cytoplasmic effectors accumulate in the biotrophic interfacial complex and move ahead of the invading pathogen, preparing rice cells for fungal colonisation.

Effector proteins generally share no common sequence features and perform different functions inside the host cell (Dangl & McDowell, 2006). A total of 851 *in planta* expressed genes encoding predicted effector proteins have been identified by a genome-wide transcriptome analysis of blast-infected rice leaves (Chen et al., 2013). Twenty *M. oryzae* effector proteins have been characterised including nine Avr effectors (Chen et al., 2013; Liu et al., 2013).

ETI is a rapid and robust plant response that is generally associated with localised cell death, known as hypersensitive response (Liu et al., 2013). It is based on the activation of highly variable resistance proteins (R proteins) that causes strong race-specific resistance. R proteins mostly contain a central nucleotide-binding site (NBS) that binds ATP or ADP, and a C-terminal leucine-rich repeats (LRRs) domain. ETI is also named the gene-for-gene interaction. Many plant NBS-LRR proteins also contain an N-terminal TIR (Toll, interleukin-1 receptor, resistance proteins) or a coiled-coil (CC) domain, whereas others have no conserved N-terminal region (Dodds & Rathjen, 2010).

NBS-LRR proteins can recognise pathogen effectors directly by physical interaction or indirectly. For example, direct binding occurs between the rice CC-NBS-LRR Pi-ta protein and the *Magnaporthe* AvrPita (Jia et al., 2000). When the pathogen recognition is indirect, the plant NBS-LRR detects either a modification of the effector's host target protein (called guardee) or a modification of a host protein that mimics the effector target (called decoy). Another possibility of indirect recognition is the "bait-and-switch" model that comprises a two-step recognition event. First, an effector interacts with the accessory "bait" protein associated with an NBS-LRR. In a subsequent step, recognition occurs between the effector and NBS-LRR protein to trigger signalling. Recent research has indicated that disease resistance or even the recognition of an AVR protein requires complementary pairs of genes (NLR genes) (Eitas & Dangl, 2010).

As pointed out before, 25 rice blast R genes were cloned and have been characterised (Devanna & Sharma, 2018). Most of them are NBS-LRR proteins except *Pi-d2* that encodes an RLK protein and Pi21 that is a proline-rich protein without any homologue in the databases. For the pathogen, only five cognate *Avr* genes have been cloned. Therefore, five pairs of *Avr* and *R* genes have been studied and both direct and indirect interactions occur between the rice R proteins and *Magnaporthe* Avr effectors (Liu et al., 2014). *M. oryzae* interaction with wheat is not as well understood as with rice. In all, ten resistance genes and one chromosome segment were identified but none of them were cloned. *AVR-Rmg8* is one of the avirulence genes cloned from *M. oryzae Triticum* pathotype (Anh et al., 2017). It encodes a small protein with 109 amino acids and a putative signal peptide but with no similarity to other known proteins. *AVR-Rmg8* is recognised by the resistance genes *Rmg7* and *Rmg8* (Table 5.1), but their molecular interaction remains unknown.

PWT3 and *PWT4* were cloned from an *Avena* isolate. Both genes encoded proteins with putative signal peptides but with no similarity to known proteins or protein domains (Inoue et al., 2017). Their sequences were compared to isolates from different hosts including wheat. *PWT4* homologues were found in some isolates but were considered non-functional whereas *PWT3* homologues were found in all isolates tested. *PWT3* and *PWT4* were employed to transform the *Triticum* isolate Br48. Inoculations with the wild type and both transformants revealed high frequencies of distribution of *Rmg3* and *Rmg4* in a world wheat collection. These resistance genes conditioned the subgroup-genus specificity rather than the race-cultivar specificity. The authors postulate that the extensive cultivation of the Anahuac variety in the 1980s in Brazil, carrying the *rwt3* gene, allowed the maintenance of *PWT3* in the *Triticum* isolates population whose loss of function makes the wheat population vulnerable to new strains. More detailed studies should be undertaken in order to understand the interaction between *Avr* genes and their cognate resistance genes.

5.5 TRANSCRIPTOMIC APPROACH

Plant diseases occur when a pathogen is able to overcome both PTI and ETI. These recognition systems involve signalling cascades to trigger plant defence responses following an intensive transcriptional reprogramming with overlap between PTI- and ETI-related differentially expressed genes (DEGs).

Few studies about transcripts modulation in the wheat–*M. oryzae Triticum* interaction have been reported (Tufan et al., 2009; Ha et al., 2016; Delventhal et al., 2017). Transcriptional changes were analysed in the leaves of wheat seedlings 24 hours post-inoculation with *Magnaporthe* sp. using the Affymetrix wheat genome array GeneChip. A large proportion of regulated probe sets was common to interactions with adapted or non-adapted isolates. Annotations of up-regulated genes are related to defence, the production of phenolic compounds and defence signalling irrespective of the interaction tested (Tufan et al., 2009). Less than 20% of the pathogen-regulated genes were differentially expressed between host and non-host interactions of wheat with *Magnaporthe* isolates. A general pathogen-induced reprogramming occurs independently of the isolate being adapted or non-adapted (Delventhal et al., 2017). Considering wheat spikes, the expression of eight resistance-related genes was analysed in genotypes differing in resistance to wheat blast and to *Fusarium* head blight. Seven of them were more responsive to *F. graminearum* than to *M. oryzae.* Specific differences in gene expression rely on wheat genotype or on pathogen rather than on a compatible or an incompatible interaction (Ha et al., 2016).

Clear differences could be noticed concerning the bleaching rate of BRS 209 and CBFusarium ENT014 (Figure 5.3). Beginning at 7 dpi, we observed a bleaching appearance in both susceptible varieties, Anahuac 75 and BRS 209. Based on this phenotype with delayed disease progress, we infer that CBFusarium ENT014 possesses specific mechanisms to restrict post-penetration events rather than the penetration of the pathogen. During the penetration phase, the cellular turgor of the appressorium of *Magnaporthe* is translated into mechanical force. This turgor pressure is equivalent to 80 atmospheres (the pressure that is felt at 800 meters under water), which is hardly endured by plant cuticle cells (Ebbole, 2007). This pressure likely explains the high number of susceptible reactions that were identified in the wheat germplasm when it was evaluated for blast disease. Moreover, the penetration process seems to depend on the biology of the pathogen regardless of the host genotype (Kim et al., 2001).

To identify the fungal effector–triggered signalling genes that contribute to slowing down pathogen colonisation, two wheat genotypes with contrasting reactions to the *M. oryzae Triticum* infection were used and RNA were isolated from spikes that were collected 40 hours post-inoculation. Cytological studies have demonstrated the initial plant cell invasions as biotrophic (Koga et al., 2004). Live-cell imaging was used to characterise the development of pathogen infectious hyphae in a cell-to-cell movement using plasmodesmata (Kankanala et al., 2007). The initial plant cell invasion is biotrophic, and when the fungus moves into another cell, the first cell loses its viability. Mosquera et al. (2009) developed a procedure to analyse genes from the pathogen and host plant contributing to biotrophic pathogen invasion. These authors used tissue that was collected 36 hpi, when infection development was relatively synchronous. In wheat, Tufan et al. (2009) observed that adapted *M. oryzae Triticum* isolates developed hyphae within epidermal and mesophyll cells with multicellular infection sites at 72 hpi. Therefore, we estimate that 40 hpi is a good time point at which to sample host tissues triggered by pathogen colonisation.

The RNA-seq strategy was adopted to examine the wheat genotypes BRS 209 and CBFusarium ENT014 and to compare the overall gene expression levels in spikes when submitted or not to inoculation with *M. oryzae Triticum* inoculum suspension using three replicates. For each of the three biological replicates, a strand-specific RNA-seq library was sequenced using Illumina HiSeq 2500 technology, producing a minimum of 20 million 2 × 125 bp paired-end reads. TMM-normalised RPKMs values were used for a digital gene expression (DGE) exploration using the edgeR package (McCarthy et al., 2012). The annotation of gene sequences and functional enrichment was done by integrating the information of several databases such as UniProt, NCBI, GO, KEGG, EggNOG and Pfam.

In this study, using a maximum false discovery rate (FDR) of 0.001, we identified 388 differentially expressed genes in BRS 209 (highly susceptible to the pathogen) and 249 DEGs in CBFusarium ENT014 (resistant to bleaching development) of which 198 and 95 with annotation, respectively. This result is quite interesting, as the susceptible reaction at 40 hpi showed a larger number of DEGs than did the resistant reaction. Considering both rice and blast fungus, more drastic changes in gene expression are observed in incompatible than in compatible interactions (Kawahara et al., 2012). These authors also examined the global expression changes by RNAseq but at 24 hpi.

The molecular functions most representative in GO terms were "heme binding", "iron ion binding" and "oxidoreductase activity, acting on paired donors, with incorporation or reduction of molecular oxygen" for both genotypes. However, they exhibited a reversal expression modulation: in BRS 209 most genes assigned to these GO terms were down-regulated, whereas in CBFusarium ENT014 most genes assigned to these GO terms were up-regulated, in response to inoculation with *M. oryzae Triticum.*

Both the heme binding and the iron ion binding functions are related with iron (Fe). The iron ion binding is the molecular function that selectively and non-covalently interacts with Fe ions (Ashburner et al., 2000; Carbon et al., 2009; The Gene Ontology Consortium, 2017). Heme binding is a molecular function that selectively and non-covalently interacts with heme (Ashburner et al., 2000; Carbon et al., 2009; The Gene Ontology Consortium, 2017).

Iron is an important element for most life forms; it is involved in important cellular processes such as electron transfer chain in respiration, photosynthesis and detoxification of ROS. However, excess iron can cause problems to the organism due to its potential to react with oxygen, producing damaging forms of reactive oxygen species. Therefore, the acquisition, utilisation and storage of iron are highly regulated at the cell and tissue level since it is beneficial or malefic to the organism (Aznar et al., 2015). Pathogens and their hosts compete for the acquisition of iron, and the mechanisms that regulate iron uptake, transport and storage have emerged as important players in the arms race between hosts and pathogens (Verbon et al., 2017).

Several studies of transcriptomic analysis report the regulation of genes related to iron when plants are submitted to biotic stresses, evidencing the important role of iron in the plant–pathogen interaction. In an expressed sequence tag (EST) library developed from wheat inoculated with *Blumeria graminis* f. sp. *tritici* (causal agent of wheat powdery mildew), a high percentage of Fe-related transcripts was observed (Liu et al., 2005). Iron homeostasis is closely related to the plant–pathogen interaction and iron is a factor associated with oxidative burst and the regulation of cereal defences. In response to a *B. graminis* f. sp. *tritici* attack, Fe^{3+} is deposited and accumulates at cell wall appositions in a redox-active form and is transported to vesicle-like bodies guided by actin polymerisation (Liu et al., 2007). In the same study, it is suggested that iron regulates gene expression through both Fe-itself and Fe-mediated oxidative stress.

Ferritin is an iron-binding protein found in a wide range of animals, plants and bacteria (Zhao, 2010). Transgenic tobacco plants ectopically expressing the iron-binding protein that promotes the accumulation of ferritin in their leaves, exhibited tolerance to necrotic damage caused by the tobacco necrosis virus and to *Alternaria alternata* and *Botrytis cinerea* fungal infections (Deák et al., 1999). The leaves of a transgenic soybean carrying an oxalate oxidase gene that degrades oxalic acid (OA), the major virulence factor of *Sclerotinia sclerotiorum* (causal agent of white mould disease), presented the expression levels of several genes altered when inoculated with OA, with ferritin showing the strongest induction (Calla et al., 2014). These studies suggest that ferritin assists in the defence against the fungi pathogen, mainly because it has an antioxidant function (Briat et al., 2010).

Heme is a cyclic tetrapyrrole (porphyrin) complex with iron at its centre (Vavilin & Vermaas, 2002). Rice (*Oryza sativa*) cv. Nipponbare plants presented a SOUL heme-binding protein down-regulated when inoculated with *Puccinia triticina* f. sp. *tritici* (Li et al., 2012). Considering the plant defence, the regulation of this protein is associated with protection against cell death and membrane disruption and may explain why necrosis or a hypersensitive response was not observed after inoculation. Rice treated with probenazole, a chemical inducer of disease resistance, induced the *OsHAP2E*, a transcription factor heme activator protein (Shimono et al., 2003). Rice lines overexpressing *OsHAP2E* showed disease indices significantly lower than the wild type (Alam et al., 2015). The suggestion is that *OsHAP2E* is involved in the salicylic acid (SA) signalling pathways since it is induced by exogenous SA.

The oxidoreductase activity, acting on paired donors, with the incorporation or reduction of molecular oxygen is defined by the catalysis of an oxidation-reduction (redox) reaction in which hydrogen or electrons are transferred from each of two donors, and molecular oxygen is reduced or incorporated into a donor (Ashburner et al., 2000; Carbon et al., 2009; The Gene Ontology Consortium, 2017). The oxidoreductases can be oxidases, dehydrogenases, peroxidases, hydroxylases, oxygenases and reductases.

Studies have identified the oxidoreductase activity in response to a fungal infection. Transcriptomic analyses of leaves of the resistant accession PI272560 (*Triticum monoccocum* var. *monococcum*) and the susceptible accession 36554 (*Triticum boeoticum* ssp. *thaoudar* var. *reuteri*) inoculated with *P. triticina* f. sp. *tritici* showed that oxidoreductase activity is the fourth molecular function most expressed in both materials (Serfling et al., 2016). Oxalate oxidase and peroxidases were identified as resistance candidate genes to diseases such as tan spot (*Pyrenophora tritici-repentis*) and stem rust (*Puccinia graminis* f. sp *tritici*), indicating that they are race non-specific and involved in the production or modulation of active oxygen species or in the oxidative burst (Faris et al., 1999).

We are focused on identifying genes potentially involved in the innate immune mechanisms of the resistant genotype against blast pathogen. Among 249 DEGs of CBFusarium ENT014, 24 are defence related and 23 are up-regulated in spikes submitted to inoculation (Table 5.5). Only one probable LRR receptor-like serine/threonine-protein kinase (TRINITY_GG_24416_c5_g1) presented a down-regulation following treatment with an inoculum suspension of *M. oryzae Triticum.*

RLKs in plants are involved in a diverse array of plant responses, including development, growth, hormone perception and response to pathogens (Goff & Ramonell, 2007). Host sensors of PAMPs are typically receptor kinases (Ronald & Beutler, 2010). In this work, we identified 15 RLKs as DEGs only in the resistant wheat genotype following inoculation. RLK-type receptors comprise the largest family of receptors in plants (Tör et al., 2009). These proteins function as a central processing unit, accepting parallel input signals internally and externally. RLKs convert input into appropriate output signals, leading to changes in metabolism, gene expression and development (Hardie, 1999). Due to their unique structural features, RLKs are suitable for cell-to-cell signalling and are considered one of the most important groups of cell surface receptors (Gou et al., 2010).

Receptor-protein kinases can be divided into subclasses according to the substrate specificities among the kinase domains: tyrosine kinases (the predominant type in animals) and serine/threonine kinases and histidine kinases, both found in plants (Afzal et al., 2008). The kinase domains of different RLKs have more than 45% identity at the amino acid level, whereas their extracellular domains have little sequence similarity (Hu et al., 2005). In this work, among the 15 RLKs identified, the serine-threonine kinase domain was found in all of them. Serine/threonine kinases are implicated in catalysing the transfer of the gamma-phosphoryl group from ATP to serine/threonine residues on protein substrates. They are known to modulate pathways that are involved in plant resistance to pathogen infection (PTI) (Yan et al., 2012).

RLKs with different extracellular motifs were identified: two DEGs coding for G-type lectin S-receptor-like serine/threonine-protein kinase (TRINITY_GG_33658_c0_g1 and TRINITY_GG_36806_c10_g3), one cysteine-rich RLK

TABLE 5.5
Defence-Related DEGs Modulated in CBFusarium ENT014 Spikes Submitted to Inoculation with *Magnaporthe oryzae Triticum* Using Three Replicates

Gene ID	log2FC	P-value	FDR	Annotation (RecName)	Gene ontology	Pfam
TRINITY_GG_97788_c1_g1	3.588679699	4.60E-25	1.19E-20	Peroxidase	GO:0005576, GO:0020037, GO:0046872, GO:0004601, GO:0042744, GO:0006979	*Peroxidase* (PF00141)
TRINITY_GG_36806_c10_g3	9.859940027	1.14E-15	7.02E-12	G-type lectin S-receptor-like serine/threonine-protein kinase SD1-1	GO:0016021, GO:0005886, GO:0009506, GO:0005524, GO:0005516, GO:0030246, GO:0004674, GO:0031625, GO:0006468, GO:0048544	–
TRINITY_GG_160722_c0_g1	3.142223315	7.34E-14	3.81E-10	Pathogenesis-related protein 1C	GO:0006952, GO:0009607	*Thaumatin* (PF00314)
TRINITY_GG_82899_c1_g1	3.438627358	1.11E-12	4.97E-09	Thaumatin-like protein PWIR2	GO:0006952, GO:0009607	*Thaumatin* (PF00314)
TRINITY_GG_24416_c5_g1	–2.273312859	1.46E-10	3.64E-07	Probable LRR receptor-like serine/threonine-protein kinase At1g51880	GO:0016021, GO:0005524, GO:0004674	–
TRINITY_GG_56385_c0_g1	3.280209878	3.33E-10	7.32E-07	Protein WIR1A	GO:0016021	–
TRINITY_GG_101753_c0_g1	2.168975032	6.47E-10	1.25E-06	Probable LRR receptor-like serine/threonine-protein kinase At3g47570	GO:0016021, GO:0005886, GO:0005524, GO:0004674	*LRRNT_2* (PF08263), *LRR_8* (PF13855), *LRR_6* (PF13516), *LRR_1* (PF00560), *Pkinase_Tyr* (PF07714), *Pkinase* (PF00069)
TRINITY_GG_104629_c0_g1	2.749445904	9.82E-10	1.82E-06	Chitinase 5	GO:0008061, GO:0004568, GO:0016998, GO:0006032, GO:0006952, GO:0000272	*Chitin_bind_1* (PF00187), *Glyco_hydro_19* (PF00182)

(Continued)

TABLE 5.5 (CONTINUED)
Defence-Related DEGs Modulated in CBFusarium ENT014 Spikes Submitted to Inoculation with *Magnaporthe oryzae Triticum* Using Three Replicates

Gene ID	log2FC	P-value	FDR	Annotation (RecName)	Gene ontology	Pfam
TRINITY_GG_107002_c3_g2	3.499080002	8.51E-09	1.21E-05	Peroxidase	GO:0005576, GO:0020037, GO:0046872, GO:0004601, GO:0042744, GO:0006979	*peroxidase* (PF00141)
TRINITY_GG_36806_c9_g1	2.824143028	5.94E-08	6.11E-05	Cysteine-rich receptor-like protein kinase 6	GO:0016021, GO:0005886, GO:0005524, GO:0004674, GO:0042742, GO:0006468	*Stress-antifung* (PF01657)
TRINITY_GG_33658_c0_g1	2.173624848	9.26E-08	8.89E-05	G-type lectin S-receptor-like serine/threonine-protein kinase At2g19130	GO:0016021, GO:0005886, GO:0005524, GO:0005516, GO:0030246, GO:0004674, GO:0006468, GO:0048544	*B_lectin* (PF01453), *S_locus_glycop* (PF00954), *PAN_2* (PF08276), *Pkinase_Tyr* (PF07714), *Pkinase* (PF00069)
TRINITY_GG_67133_c0_g1	8.933376849	1.75E-07	0.000149433	Wall-associated receptor kinase-like 1	GO:0016021, GO:0009505, GO:0005886, GO:0005524, GO:0030247, GO:0004674, GO:0007166	–
TRINITY_GG_20175_c21_g1	1.760198418	1.82E-07	0.000151301	Basic endochitinase A	GO:0005576, GO:0008061, GO:0004568, GO:0016998, GO:0006032, GO:0050832, GO:0031640, GO:0000272	*Chitin_bind_1* (PF00187), *Glyco_hydro_19* (PF00182)
TRINITY_GG_127434_c2_g1	2.385593715	1.91E-07	0.000156096	MDIS1-interacting receptor like kinase 2	GO:0016021, GO:0016020, GO:0005886, GO:0009506, GO:0090406, GO:0005524, GO:0004674	*LRR_1* (PF00560), *LRR_8* (PF13855)

(Continued)

TABLE 5.5 (CONTINUED)
Defence-Related DEGs Modulated in CBFusarium ENT014 Spikes Submitted to Inoculation with *Magnaporthe oryzae Triticum* Using Three Replicates

Gene ID	log2FC	P-value	FDR	Annotation (RecName)	Gene ontology	Pfam
TRINITY_GG_28163_c15_g1	2.656635657	1.95E-07	0.000157714	LRR receptor kinase SERK2	GO:0016021, GO:0005886, GO:0005524, GO:0042803, GO:0004672, GO:0004675, GO:0009742, GO:0007166, GO:0009729, GO:0006468	*LRRNT_2* (PF08263), *LRR_1* (PF00560), *LRR_8* (PF13855)
TRINITY_GG_20131_c4_g1	8.869539194	2.07E-07	0.000165649	Probable LRR receptor-like serine/threonine-protein kinase At3g47570	GO:0016021, GO:0005886, GO:0005524, GO:0004674	–
TRINITY_GG_51766_c1_g1	2.643866224	2.39E-07	0.000187856	Protein WIR1A	GO:0016021	–
TRINITY_GG_86562_c0_g1	1.453396001	1.05E-06	0.000626716	Protein SUPPRESSOR OF NPR1-1 CONSTITUTIVE 4	GO:0016021, GO:0005886, GO:0005524, GO:0008889, GO:0004674, GO:0006968, GO:0006952, GO:0006071, GO:0045087, GO:0006629	*Pkinase* (PF00069), *Pkinase_Tyr* (PF07714)
TRINITY_GG_122913_c0_g1	1.880865297	1.09E-06	0.000642849	Glucan endo-1,3-beta-glucosidase GII	GO:0042973, GO:0006076, GO:0050832	*Glyco_hydro_17* (PF00332)
TRINITY_GG_67155_c0_g1	2.416215721	1.38E-06	0.000768284	Wall-associated receptor kinase 3	GO:0016021, GO:0005524, GO:0005509, GO:0030247, GO:0004674, GO:0007166	*Pkinase* (PF00069), *Pkinase_Tyr* (PF07714)
TRINITY_GG_106713_c19_g1	8.746158257	1.57E-06	0.000854893	Putative leucine-rich repeat receptor-like serine/threonine-protein kinase At2g24130	GO:0016021, GO:0005886, GO:0005524, GO:0004674	–

(Continued)

TABLE 5.5 (CONTINUED)
Defence-Related DEGs Modulated in CBFusarium ENT014 Spikes Submitted to Inoculation with *Magnaporthe oryzae Triticum* Using Three Replicates

Gene ID	log2FC	P-value	FDR	Annotation (RecName)	Gene ontology	Pfam
TRINITY_GG_70444_c0_g1	1.691064186	1.61E-06	0.000870441	Probable leucine-rich repeat receptor-like protein kinase At1g35710	GO:0016021, GO:0016020, GO:0005886, GO:0009506, GO:0090406, GO:0005524, GO:0004674	*LRR_8* (PF13855), *LRR_1* (PF00560)
TRINITY_GG_9503_c2_g1	2.21110391	1.65E-06	0.000882886	Probable LRR receptor-like serine/threonine-protein kinase At1g67720	GO:0016021, GO:0005886, GO:0005524, GO:0004674, GO:0042742, GO:0012501, GO:0006468, GO:0009751	Pkinase (PF00069), *Pkinase_Tyr* (PF07714), *Jacalin* (PF01419)
TRINITY_GG_122631_c0_g1	2.572076151	1.82E-06	0.000956253	LRR receptor-like serine/threonine-protein kinase FLS2	GO:0016021, GO:0005886, GO:0005524, GO:0004674	*LRR_8* (PF13855), *LRR_1* (PF00560), *LRR_6* (PF13516), *Pkinase* (PF00069), *Pkinase_Tyr* (PF07714)

encoding gene (TRINITY_GG_36806_c9_g1) and one DEG coding for an atypical receptor-like kinase (Protein SUPPRESSOR OF NPR1-1 CONSTITUTIVE 4-SNC4, TRINITY_GG_86562_c0_g1). Lectin receptor kinases (LecRK) present a lectin motif extracellularly and they are classified into four types: G, C, L and LysM type (Lannoo & Van Damme, 2014). Lectins are soluble or membrane proteins found in a wide range of organisms. They are present inside and outside cells and have a carbohydrate-binding site. G-type LecRKs function in self-incompatibility reactions in flowering plants and in plant defence (Singh & Zimmerli, 2013). In rice, 100 G-type LecRKs have been identified (Vaid et al., 2012). *Pi-d2*, one of the rice genes conferring resistance to *M. oryzae*, is a G-type LecRK (Chen et al., 2006). Cysteine-rich, receptor-like kinases (CRKs) present one or more repeats of the "domain of unknown function 26" (DUF26) consisting of a C-X8-C-X2-C motif in the extracellular part of a protein. Members of this family are induced by salicylic acid and pathogens and some of them are also involved in the regulation of cell death in *Arabidopsis* (Chen et al., 2004). *TaCRK1* is the first CRK gene isolated from wheat. Higher expression levels were observed following the inoculation of *Rhizoctonia cerealis* in resistant wheat and after ABA exogenous application. The TaCRK1 protein may localise to plasma membranes (Yang et al., 2013). SNC4 is a receptor-like kinase with two putative extracellular glycerophosphoryl diester phosphodiesterase domains. In *Arabidopsis*, the *snc4-1D* mutant had a constitutive expression of defence marker genes such as *PR1*, *PR2* and *PDF1.2*, accumulating high levels of hydrogen peroxide and presenting microscopic cell death (Bi et al., 2010). It is hypothesised that a PAMP receptor activates multiple downstream resistance pathways.

Nine RLKs contain leucine-rich repeats (LRRs) in the extracellular region. Members of this subfamily are involved in both plant signalling initiation and activation processes. They are the largest and the best-studied RLK subfamily (Peng et al., 2009). The rice genome has been shown to contain around 320 LRR-RLK genes (Dievart et al., 2016). They are involved in responses to diverse signals such as hormones, steroids, light, wounding and pathogens. Some examples of LRR-RLKs are XA21 that confers broad-spectrum resistance to diverse strains of *Xanthomonoas oryzae* pv. *oryzae* (Xoo); BRASSINOSTEROID INSENSITIVE 1 (BRI1) involved in the perception of brassinosteroid; CLAVATA1 related to the control of meristem cell fate; PEP RECEPTOR 1 (PEPR1) and PEPR2 that sense the damage-associated molecular pattern peptide 1 (Pep1); two important PRRs, FLAGELLIN SENSING2 (FLS2) and EF-TU receptor (EFR), that recognise bacterial flagellin and EF-Tu, respectively; and the SERK subfamily (Chen & Ronald, 2011; Yeh et al., 2015). *OsSERK1* mediates somatic embryogenesis in rice and is induced early by an avirulent blast isolate and by defence signalling molecules such as salicylic acid and jasmonic acid. Sekigushi lesion mimic mutants presented a constitutive expression of *OsSERK1* indicating its possible implication in host cell death (Hu et al., 2005). Also in rice, *OsBRR1* (blast resistance related) encodes an LRR-RLK whose expression is highly induced in leaves infected by *M. oryzae*. Suppression and overexpression of *OsBRR1* led to increased susceptibility and enhanced resistance to the pathogen, respectively (Peng et al., 2009).

Two DEGs (TRINITY_GG_67133_c0_g1 and TRINITY_GG_67155_c0_g1) encoding wall-associated kinases (WAKs) were identified. WAKs are a subfamily of plant RLKs. They possess an amino-terminal domain closely associated with the extracellular matrix, a single transmembrane-spanning region and a cytoplasmic kinase domain (He et al., 1998). WAK-type RLKs are associated with pectin, which is present in the middle lamella and primary cell wall (Wagner & Kohorn, 2001). Plant cell walls are composed of a complex array of cellulose, hemicellulose, pectins and proteins that surround each cell. WAK proteins are required for cell expansion and are also involved in the response to pathogens. In fact, pectin fragments are commonly generated by pathogens and can serve as signals to activate specific signalling pathways related to plant defence (Kohorn & Kohorn, 2012). Although it is not clear how WAKs are involved, they are known to be associated with pathogenesis (Zhang et al., 2005). In rice, a WAK-like (WAKL) allele is less sensitive to rice blast disease (Li et al., 2009).

The most effective plant defence is the hypersensitive reaction that restricts the growth and damage caused by pathogens. Altered ion fluxes across the plant cell membrane, generation of active oxygen species, changes in the phosphorylation state of regulatory proteins and transcriptional activation of defence-related genes take place in response to pathogen infection. Cell death occurs at the site of infection and is accompanied by a process of lignification, suberisation and deposition of callose to strengthen the cell wall and the production of phenolic compounds, phytoalexins and PR proteins. PR proteins accumulate locally in the infected and surrounding tissues, and also in remote uninfected tissues (Ebrahim et al., 2011). In the latter case, they are associated with the development of systemic acquired resistance (SAR), which enhances the plant's capacity to defend against specially biotrophic and hemibiotrophic pathogens (Hammerschmidt, 2009).

PR proteins have been classified into 17 families of which some are antimicrobial. In our study, we identified three DEG sequences coding for proteins involved in the degradation of components of the fungal cell wall, one glucan endo-1,3-beta-glucosidase (TRINITY_GG_122913_c0_g1) and two chitinases (TRINITY_GG_104629_c0_g1 and TRINITY_GG_20175_c21_g1). After cellulose, chitin is the most abundant polymer in living organisms. It is the major component of the fungal cell wall. Chitinases are part of the group of pathogenesis-related (PR) proteins and are strongly induced in plant cells infected by pathogens (Cletus et al., 2013). A class VII chitinase isolated and purified from wheat exerted a broad-spectrum antifungal activity against several pathogens including *Fusarium* sp., causal agent of scab (Singh et al., 2007). Beta 1,3-glucanase belongs to O-Glycosyl hydrolases that hydrolyse the glycosidic bond between two or more carbohydrates, or between a carbohydrate and a non-carbohydrate moiety. Together with lichenases and exo-1,3-glucanase, endo-1,3-beta-glucanase are currently found in plants and in fungi. A coordinated expression of beta 1,3-glucanases and chitinases has been described in many plant species including wheat spikes infected by *Fusarium graminearum* (Li et al., 2001). Transgenic plants expressing a beta 1,3-glucanase and chitinase gene combination present enhanced fungal disease resistance and delayed spread of the *F. graminearum* infection in wheat spikes compared to control plants (Anand et al., 2003).

Besides chitinases and beta 1,3-glucanase, two DEG sequences (TRINITY_GG_160722_c0_g1 and TRINITY_GG_82899_c1_g1) coding for thaumatin-like proteins (TLPs) were identified. Coincident with these results, a beta 1,3-glucanase, a chitinase and a thaumatin-like protein were up-regulated in wheat spikes infected by *M. oryzae* (Ha et al., 2016). TLPs contain around 200 amino acid residues and are universal in plants, including bryophytes, gymnosperms and angiosperms (Liu et al., 2010b). Their expression is inducible by pathogen attack and they are classified as PR protein family 5. In this family, there are TLPs induced by environmental stresses including drought, cold, salt, wounding and plant hormones. Other plant TLPs have been reported as food allergens from fruits and pollen allergens from conifers (Breiteneder, 2004). TLPs present a well-documented antifungal activity that could be related to its binding and/or hydrolysing activity to water-insoluble beta 1,3-glucans, an important component of the fungal cell wall. In contrast, some TLPs do not show antifungal activity despite their glucanase activity. In fact, they could possibly interfere with the fate and metabolism of plant-derived beta 1,3-glucans as a consequence of wounding or callose deposition (Grenier et al., 1999). Xylanase inhibition is another mechanism of TLPs that act in plant defence. Xylan is one of the most abundant polysaccharides in the cell wall of higher plants. Xylanases can be produced by microorganisms and by plants. The recently discovered wheat thaumatin-like xylanase inhibitor presents activity exclusively against xylanases of microbial origin (Dornez et al., 2010).

Peroxidases are enzymes that catalyse oxidoreduction between H_2O_2 and several reductants. We identified two DEG sequences of peroxidases in our study (TRINITY_GG_97788_c1_g1 and TRINITY_GG_107002_c3_g2). Both are class III peroxidase, a classical enzyme with a heme-binding domain, encoded by a large multigene family in plants. Among other processes, class III peroxidases are induced during the host plant defence. Peroxidases are also known to belong to the PR protein subfamily 9 (van Loon et al., 2006). In the plant-pathogen defence, the peroxidases act in the reinforcement of the cell walls (e.g. suberization, lignification), in the metabolism of ROS and reactive nitrogen species, in the production of antimicrobial metabolites and in the signalling pathways (Hiraga et al., 2001; Almagro et al., 2009). The induction of peroxidases in response to *M. oryzae* infection has been reported in several studies. Rice plants at the eight-leaf stage presented transcripts accumulation of eight peroxidase genes after *M. oryzae* inoculation (Sasaki et al., 2004). Some of them were detected at 6–12 hours post-inoculation, when the symptoms are not visible and others at 5 days after inoculation, when the lesions are visible. A rice-resistant line exhibited broad-spectrum changes in gene regulation in comparison to susceptible line, when inoculated with *M. oryzae*, with very high up-regulation of peroxidases genes (Gupta et al., 2012).

Finally, two DEGs (TRINITY_GG_56385_c0_g1 and TRINITY_GG_51766_c1_g1) encoding for Wheat Induced Resistance 1 (WIR1) proteins were identified. The *WIR1* is a gene family supposedly related to basal host resistance. This gene family was first identified in wheat (cv. Fidel) inoculated with a barley isolate of *B. graminis* f. sp. *hordei* (Schweizer et al., 1989), and is composed by three sequences: *TaWIR1a*, *TaWIR1b* and *TaWIR1c* (Bull et al., 1992; Franck & Dudler, 1995). Homologues of *TaWIR1* have been found in rice (Mauch et al., 1998) and in barley (Gaupels et al., 2008; Douchkov et al., 2011). *TaWIR1* encodes for an integral membrane protein, with a polar N-terminal domain pointing to the cytoplasm and an extracytoplasmic C-terminal domain rich in glycine and proline (Bull et al., 1992). In wheat, the accumulation of *WIR1* transcripts has been shown in response to the biotrophic pathogens *B. graminis* f. sp. *hordei* (Schweizer et al., 1989), *P. triticina* (Bolton et al., 2008) and *P. striiformis* f. sp. *tritici* (Coram et al., 2007; Bozkurt et al., 2010), and to the hemibiotrophic pathogens *F. pseudograminearum* (Desmond et al., 2008), *M. oryzae.* (Tufan et al., 2009) and *F. graminearum* (Diethelm et al., 2012). In the wheat–*M. oryzae* interaction, the silencing of *TaWIR1* had no effect on the initial penetration of *M. oryzae* in epidermal cells of wheat seedlings (Tufan et al., 2012). However, after the establishment of the infection site, the intracellular spread of the fungus was facilitated. It has been suggested that the TaWIR1 proteins may be related to increased adhesion of the plasma membrane to the cell wall during a pathogen attack, which is required for expression of plant defense responses during fungal penetration (Bull et al., 1992).

5.6 FUTURE PROSPECTS

Many aspects of the wheat–*M. oryzae Triticum* interaction remain to be elucidated. Genetic mapping has been efficient in identifying genes and QTL related to blast resistance in rice. The development of "pop Embrapa-brusone", a blast-specific mapping population, will allow to clarify of the genetic basis of wheat resistance and grain yield upon infection conditions.

Candidate genes identified as differentially expressed in CBFusarium ENT014 in response to *Magnaporthe* infection present unique features for pathogen sensing and for downstream immunity signalling. Additional studies are required to analyse in detail their expression and their roles in the slowed development of bleaching symptoms. The information coming from both approaches in concert should support the development of molecular markers to assist breeding for blast resistance.

ACKNOWLEDGEMENTS

Research was supported by Embrapa projects: Wheat BGIn (SEG 02.08.01.006.00.00) and Wheat BGIn2 (SEG 02.11.08.004.00.00). J. R. Ferreira and G. A. Camilotti were supported by CAPES (edital Capes-Embrapa 15/2014) and PIBIC-CNPq (proc 45125/2016-5) fellowships, respectively. The authors have no conflicts of interest to declare.

REFERENCES

Afzal, A. J., A. J. Wood and D. A. Lightfoot. "Plant Receptor-Like Serine Threonine Kinases: Roles in Signaling and Plant Defense." *Molecular Plant-Microbe Interactions* 21, no. 5 (2008): 507–17.

Alam, M. M., T. Tanaka, H. Nakamura et al. "Overexpression of a Rice Heme Activator Protein Gene (OsHAP2E) Confers Resistance to Pathogens, Salinity and Drought, and Increases Photosynthesis and Tiller Number." *Plant Biotechnology Journal* 13, no. 1 (2015): 85–96. doi:10.1111/pbi.12239.

Almagro, L., L. V. Gómez Ros, S. Belchi-Navarro, R. Bru, A. Ros Barceló and M. A. Pedreño. "Class III Peroxidases in Plant Defence Reactions." *Journal of Experimental Botany* 60, no. 2 (2009): 377–90. doi:10.1093/jxb/ern277.

Anand, A., T. Zhou, H. N. Trick, B. S. Gill, W. W. Bockus and S. Muthukrishnan. "Greenhouse and Field Testing of Transgenic Wheat Plants Stably Expressing Genes for Thaumatin-Like Protein, Chitinase and Glucanase Against *Fusarium graminearum*." *Journal of Experimental Botany* 54, no. 384 (2003): 1101–11. doi:10.1093/jxb/erg110.

Anh, V. L., N. T. Anh, A. G. Tagle et al. "*Rmg8*, a New Gene for Resistance to Triticum Isolates of *Pyricularia oryzae* in Hexaploid Wheat." *Phytopathology* 105, no. 12 (2015): 1568–72.

Anh, V. L., Y. Inoue, S. Asuke et al. "*Rmg8* and *Rmg7*, Wheat Genes for Resistance to the Wheat Blast Fungus, Recognize the Same Avirulence Gene AVR-Rmg8." *Molecular Plant Pathology* 19, no. 5 (2017): 1252–6. doi:10.1111/mpp.12609.

Anjos, J. R. N., D. B. da Silva, M. J. D. Charchar and G. C. Rodrigues. "Ocorrência de brusone (*Pyricularia grisea*) em trigo e centeio na região dos cerrados do Brasil Central." *Pesquisa Agropecuária Brasileira* 31, no. 1 (1996): 79–82.

Ashburner, M., C. A. Ball, J. A. Blake et al. "Gene Ontology: Tool for the Unification of Biology." *Nature Genetics* 25, no. 1 (2000): 25–9. doi:10.1038/75556.

Ashkani, S., M. Y. Rafii, M. Shabanimofrad, A. Ghasemzadeh, S. A. Ravanfar and M. A. Latif. "Molecular Progress on the Mapping and Cloning of Functional Genes for Blast Disease in Rice (*Oryza sativa* L.): Current Status and Future Considerations." *Critical Reviews in Biotechnology* 36, no. 2 (2016): 353–67.

Aznar, A., N. W. G. Chen, S. Thomine and A. Dellagi. "Immunity to Plant Pathogens and Iron Homeostasis." *Plant Science* 240 (2015): 90–7. doi:10.1016/j.plantsci.2015.08.022.

Barea, G. and J. Toledo. *Identificación y zonificación de Piricularia (Pyricularia oryzae) en el cutivo de trigo en el departamento de Santa Cruz*. Santa Cruz de la Sierra, Bolivia: Centro de Investigación Agrícola Tropical. Informe Tecnico. Proyecto de Investigación Trigo, 1996.

Bi, D., Y. T. Cheng, X. Li and Y. Zhang. "Activation of Plant Immune Responses by a Gain-of-Function Mutation in an Atypical Receptor-Like Kinase." *Plant Physiology* 153, no. 4 (2010): 1771–9. doi:10.1104/pp.110.158501.

Bolton, M. D., J. A. Kolmer, W. W. Xu and D. F. Garvin. "Lr34-Mediated Leaf Rust Resistance in Wheat: Transcript Profiling Reveals a High Energetic Demand Supported by Transient Recruitment of Multiple Metabolic Pathways." *Molecular Plant-Microbe Interactions* 21, no. 12 (2008): 1515–27. doi:10.1094/MPMI-21-12-1515.

Bonman, J. M., T. I. Vergel de Dios and M. M. Khin. "Physiologic Specialization of *Pyricularia oryzae* in the Philippines." *Plant Disease* 70, no. 8 (1986): 767–9.

Bozkurt, T. O., G. R. D. McGrann, R. MacCormack, L. A. Boyd and M. S. Akkaya. "Cellular and Transcriptional Responses of Wheat During Compatible and Incompatible Race-Specific Interactions with *Puccinia striiformis* f. sp. *tritici*." *Molecular Plant Pathology* 11, no. 5 (2010): 625–40. doi:10.1111/j.1364-3703.2010.00633.x.

Breiteneder, H. "Thaumatin-Like Proteins – A New Family of Pollen and Fruit Allergens." *Allergy* 59, no. 5 (2004): 479–81. doi:10.1046/j.1398-9995.2003.00421.x.

Briat, J.-F., K. Ravet, N. Arnaud et al. "New Insights into Ferritin Synthesis and Function Highlight a Link Between Iron Homeostasis and Oxidative Stress in Plants." *Annals of Botany* 105, no. 5 (2010): 811–22. doi:10.1093/aob/mcp128.

Bull, J., F. Mauch, C. Hertig, G. Rebmann and R. Dudler. "Sequence and Expression of a Wheat Gene that Encodes a Novel Protein Associated with Pathogen Defense." *Molecular Plant-Microbe Interactions* 5 (1992): 516–19.

Cabrera, M. G. and S. A. Gutiérrez. "Primer registro de *Pyricularia grisea* en cultivos de trigo del NE de Argentina." In *Jornada de Actualización en Enfermedades de Trigo*, 60. Lavallol, Buenos Aires: IFSC Press, 2007.

Calla, B., L. Blahut-Beatty, L. Koziol, D. H. Simmonds and S. J. Clough. "Transcriptome Analyses Suggest a Disturbance of Iron Homeostasis in Soybean Leaves During White Mould Disease Establishment." *Molecular Plant Pathology* 15, no. 6 (2014): 576–88. doi:10.1111/mpp.12113.

Carbon, S., A. Ireland, C. J. Mungall et al. "AmiGO: Online Access to Ontology and Annotation Data." *Bioinformatics* 25, no. 2 (2009): 288–9. doi:10.1093/bioinformatics/btn615.

Cardoso, C. A. A., E. M. Reis and E. N. Moreira. "Development of a Warning System for Wheat Blast Caused by *Pyricularia grisea*." *Summa Phytopathologica* 34, no. 3 (2008): 216–21.

Cesari, S., M. Bernoux, P. Moncuquet, T. Kroj and P. Dodds. "A Novel Conserved Mechanism for Plant NLR Protein Pairs: The 'Integrated Decoy' Hypothesis." *Frontiers in Plant Science* 5 (2014): 606. doi:10.3389/fpls.2014.00606.

Chen, K., B. Fan, L. Du and Z. Chen. "Activation of Hypersensitive Cell Death by Pathogen-Induced Receptor-Like Protein Kinases from Arabidopsis." *Plant Molecular Biology* 56, no. 2 (2004): 271–83.

Chen, S., P. Songkumarn, R. C. Venu et al. "Identification and Characterization of in Planta-Expressed Secreted Effector Proteins from *Magnaporthe oryzae* that Induce Cell Death in Rice." *Molecular Plant-Microbe Interactions* 26, no. 2 (2013): 191–202.

Chen, X. and P. C. Ronald. "Innate Immunity in Rice." *Trends in Plant Science* 16, no. 8 (2011): 451–9.

Chen, X., J. Shang, D. Chen et al. "A B-Lectin Receptor Kinase Gene Conferring Rice Blast Resistance." *The Plant Journal* 46, no. 5 (2006): 794–804. doi:10.1111/j.1365-313X.2006.02739.x.

Chiu, W. F. and Y. H. Chang. "Advances of Science of Plant Protection in the People's Republic of China." *Annual Review of Phytopathology* 20, no. 1 (1982): 71–92.

Choi, J., S.-Y. Park, B.-R. Kim et al. "Comparative Analysis of Pathogenicity and Phylogenetic Relationship in *Magnaporthe grisea* Species Complex." *PLOS ONE* 8, no. 2 (2013): e57196. doi:10.1371/journal.pone.0057196.

Cletus, J., V. Balasubramanian, D. Vashisht and N. Sakthivel. "Transgenic Expression of Plant Chitinases to Enhance Disease Resistance." *Biotechnology Letters* 35, no. 11 (2013): 1719–32. doi:10.1007/s10529-013-1269-4.

Coram, T. E., M. Wang and X. Chen. "Transcriptome Analysis of the Wheat–*Puccinia striiformis* f. sp. *tritici* Interaction." *Molecular Plant Pathology* 9, no. 2 (2007): 157–69. doi:10.1111/j.1364-3703.2007.00453.x.

Cruz, C. D., G. L. Peterson, W. W. Bockus et al. "The 2NS Translocation from *Aegilops ventricosa* Confers Resistance to the Triticum Pathotype of *Magnaporthe oryzae*." *Crop Science* 56, no. 3 (2016): 990–1000.

Cumagun, C. J. R., V. L. Anh, T. T. P. Vy et al. "Identification of a Hidden Resistance Gene in Tetraploid Wheat Using Laboratory Strains of *Pyricularia oryzae* Produced by Backcrossing." *Phytopathology* 104, no. 6 (2014): 634–40.

Dangl, J. L. and J. M. McDowell. "Two Modes of Pathogen Recognition by Plants." *Proceedings of the National Academy of Sciences* 103, no. 23 (2006): 8575–6.

Deák, M., G. V. Horváth, S. Davletova et al. "Plants Ectopically Expressing the Iron-binding Protein, Ferritin, are Tolerant to Oxidative Damage and Pathogens." *Nature Biotechnology* 17 (1999): 192–6. doi:10.1038/6198.

DeFeyter, R., Y. Yang and D. W. Gabriel. "Gene-for-Genes Interactions Between Cotton R Genes and *Xanthomonas campestris* pv. *malvacearum* avr Genes." *Molecular Plant-Microbe Interactions* 6, no. 2 (1993): 225–37.

Delventhal, R., J. Rajaraman, F. L. Stefanato et al. "A Comparative Analysis of Nonhost Resistance Across the two Triticeae Crop Species Wheat and Barley." *BMC Plant Biology* 17, no. 1 (2017): 232.

Desmond, O. J., J. M. Manners, P. M. Schenk, D. J. Maclean and K. Kazan. "Gene Expression Analysis of the Wheat Response to Infection by *Fusarium pseudograminearum.*" *Physiological and Molecular Plant Pathology* 73, no. 1 (2008): 40–7. doi:10.1016/j.pmpp.2008.12.001.

Devanna, B. N. and T. R. Sharma. "Wheat Blast Disease Management: Cues from the Advancements in Molecular Biology of Rice-*Magnaporthe* Pathosystem." *Journal of Plant Biochemistry and Biotechnology* 27, no. 3 (2018): 249–59. doi:10.1007/s13562-017-0435-5.

Diethelm, M., M. Rhiel, C. Wagner et al. "Gene Expression Analysis of Four WIR1-Like Genes in Floret Tissues of European Winter Wheat after Challenge with *G. zeae. Euphytica* 186, no. 1 (2012): 103–14. doi:10.1007/s10681-011-0498-7.

Dievart, A., C. Perin, J. Hirsch et al. "The Phenome Analysis of Mutant Alleles in Leucine-Rich Repeat Receptor-Like Kinase Genes in Rice Reveals New Potential Targets for Stress Tolerant Cereals." *Plant Science* 242 (2016): 240–9. doi:10.1016/j.plantsci.2015.06.019.

Dodds, P. N. and J. P. Rathjen. "Plant Immunity: Towards an Integrated View of Plant–Pathogen Interactions." *Nature Reviews Genetics* 11, no. 8 (2010): 539.

Dornez, E., E. Croes, K. Gebruers et al. "Accumulated Evidence Substantiates a Role for Three Classes of Wheat Xylanase Inhibitors in Plant Defense." *Critical Reviews in Plant Sciences* 29, no. 4 (2010): 244–64. doi:10.1080/07352689.2010.487780.

Douchkov, D., A. Johrde, D. Nowara et al. "Convergent Evidence for a Role of WIR1 Proteins During the Interaction of Barley with the Powdery Mildew Fungus *Blumeria graminis.*" *Journal of Plant Physiology* 168, no. 1 (2011): 20–9. doi:10.1016/j.jplph.2010.07.004.

Ebbole, D. J. "Magnaporthe as a Model for Understanding Host–Pathogen Interactions." *Annual Review of Phytopathology* 45 (2007): 437–56.

Ebrahim, S., K. Usha and B. Singh. "Pathogenesis Related (PR) Proteins in Plant Defense Mechanism." *Science Against Microbial Pathogens* 2 (2011): 1043–54.

Eitas, T. K. and J. L. Dangl. "NB-LRR Proteins: Pairs, Pieces, Perception, Partners, and Pathways." *Current Opinion in Plant Biology* 13, no. 4 (2010): 472–77.

Faris, J. D., W. L. Li, D. J. Liu, P. D. Chen and B. S. Gill. "Candidate Gene Analysis of Quantitative Disease Resistance in Wheat." *Theoretical and Applied Genetics* 98, no. 2 (1999): 219–25. doi:10.1007/s001220051061.

Farman, M., G. Peterson, L. Chen et al. "The Lolium Pathotype of *Magnaporthe oryzae* Recovered from a Single Blasted Wheat Plant in the United States." *Plant Disease* 101, no. 5 (2017): 684–92.

Ferreira, J. R., C. Vancini, C. C. Deuner et al. *Absence of 2NS/2AS in wheat resistance sources to Magnaporthe oryzae in Brazil, Boletim de Pesquisa e Desenvolvimento*, 91. Passo Fundo: Embrapa Trigo, 2018.

Flor, H. H. "The Complementary Genic Systems in Flax and Flax Rust." In *Advances in Genetics*, edited by M. Demerec, 29–54. Academic Press Inc. Published by Elsevier Inc. New York, NY, 1956. DOI: 10.1016/S0065-2660(08)60498-8

Franck, S. and R. Dudler. "Nucleotide Sequence (Genbank/EMBL/DDBJ Accession Number X87686) of a Wheat cDNA Encoding a Putative Pathogen-Inducible Protein Homologous to PWIR1." *Plant Physiology* 109 (1995): 338.

Gaupels, F., A. Buhtz, T. Knauer et al. "Adaptation of Aphid Stylectomy for Analyses of Proteins and mRNAs in Barley Phloem sap." *Journal of Experimental Botany* 59, no. 12 (2008): 3297–306. doi:10.1093/jxb/ern181.

Giraldo, M. C., Y. F. Dagdas, Y. K. Gupta et al. "Two Distinct Secretion Systems Facilitate Tissue Invasion by the Rice Blast Fungus *Magnaporthe oryzae.*" *Nature Communications* 4 (2013): 1996. doi:10.1038/ncomms2996.

Goff, K. E. and K. M. Ramonell. "The Role and Regulation of Receptor-Like Kinases in Plant Defense." *Gene Regulation and Systems Biology* 1 (2007): 167–75. doi:10.1177/117762500700100015.

Gou, X., K. He, H. Yang et al. "Genome-Wide Cloning and Sequence Analysis of Leucine-Rich Repeat Receptor-Like Protein Kinase Genes in *Arabidopsis thaliana*." *BMC Genomics* 11, no. 1 (2010): 19.

Goulart, A. C. P., F. A. Paiva and A. N. de Mesquita. "Ocorrência da brusone (*Pyricularia oryzae*) do trigo (*Triticum aestivum*) em Mato Grosso do Sul." *Fitopatologia Brasileira* 15, no. 1 (1990): 112–14.

Goulart, A., C. Pereira, P. G. Sousa and A. S. Urashima. "Danos em trigo causados pela infecção de *Pyricularia grisea*." *Summa Phytopathologica* 33, no. 4 (2007): 358–63.

Grenier, J., C. Potvin, J. Trudel and A. Asselin. "Some Thaumatin-Like Proteins Hydrolyse Polymeric β-1,3-glucans." *The Plant Journal* 19, no. 4 (1999): 473–80. doi:10.1046/j.1365-313X.1999.00551.x.

Gupta, S. K., A. K. Rai, S. S. Kanwar, D. Chand, N. K. Singh and T. R. Sharma. "The Single Functional Blast Resistance Gene Pi54 Activates a Complex Defence Mechanism in Rice." *Journal of Experimental Botany* 63, no. 2 (2012): 757–72. doi10.1093/jxb/err297.

Ha, X., B. Koopmann and A. Tiedemann. "Wheat Blast and Fusarium Head Blight Display Contrasting Interaction Patterns on Ears of Wheat Genotypes Differing in Resistance." *Phytopathology* 106, no. 3 (2016): 270–81.

Hammerschmidt, R. "Systemic Acquired Resistance." *Advances in Botanical Research* 51 (2009): 173–222.

Hardie, D. G. "Plant Protein Serine/Threonine Kinases: Classification and Functions." *Annual Review of Plant Biology* 50, no. 1 (1999): 97–131.

Hau, V. T., K. Bich, J. Hirata, H. Murakami, S. Mayama Nakayashiki and Y. Tosa. "Rwt4, a Wheat Gene for Resistance to Avena Isolates of *Magnaporthe oryzae*, Functions as a Gene for Resistance to Panicum Isolates in Japan." *Journal of General Plant Pathology* 73, no. 1 (2007): 22–8.

He, Z.-H., D. He and B. D. Kohorn. "Requirement for the Induced Expression of a Cell Wall Associated Receptor Kinase for Survival During the Pathogen Response." *The Plant Journal* 14, no. 1 (1998): 55–63. doi:10.1046/j.1365-313X.1998.00092.x.

Hiraga, S., K. Sasaki, H. Ito, Y. Ohashi and H. Matsui. "A Large Family of Class III Plant Peroxidases." *Plant and Cell Physiology* 42, no. 5 (2001): 462–68. doi:10.1093/pcp/pce061.

Hirata, K., Y. Tosa, H. Nakayashiki and S. Mayama. "Significance of PWT4–Rwt4 Interaction in the Species Specificity of Avena Isolates of *Magnaporthe oryzae* on Wheat." *Journal of General Plant Pathology* 71, no. 5 (2005): 340–4. doi:10.1007/s10327-005-0215-2.

Hu, H., L. Xiong and Y. Yang. "Rice SERK1 Gene Positively Regulates Somatic Embryogenesis of Cultured Cell and Host Defense Response Against Fungal Infection." *Planta* 222, no. 1 (2005): 107–17. doi:10.1007/s00425-005-1534-4.

Igarashi, S. "Update on Wheat Blast (*Pyricularia oryzae*) in Brazil." In *Wheat for the Nontraditional Warm Areas*, edited by D. A. Saunders, 480–3. Mexico, DF: CIMMYT, 1991.

Igarashi, S., C. M. Utiamada, L. C. Igarashi, A. H. Kazuma and R. S. Lopes. "*Pyricularia em trigo*. 1. Ocorrência de Pyricularia sp. no estado do Paraná." *Fitopatologia Brasileira* 11, 1986: 351–2.

Inoue, Y., Trinh, T. P. Vy, K. Yoshida et al. "Evolution of the Wheat Blast Fungus Through Functional Losses in a Host Specificity Determinant." *Science* 357, no. 6346 (2017): 80–3.

Jia, Y., S. A. McAdams, G. T. Bryan, H. P. Hershey and B. Valent. "Direct Interaction of Resistance Gene and Avirulence Gene Products Confers Rice Blast Resistance." *The EMBO Journal* 19, no. 15 (2000): 4004–14.

Jones, J. D. G. and J. L. Dangl. "The Plant Immune System." *Nature* 444, no. 7117 (2006): 323–9.

Kang, S., J. A. Sweigard and B. Valent. "The PWL Host Specificity Gene Family in the Blast Fungus *Magnaporthe grisea*." *Molecular Plant-Microbe Interactions* 8, no. 6 (1995): 939–48.

Kankanala, P., K. Czymmek and B. Valent. "Roles for Rice Membrane Dynamics and Plasmodesmata During Biotrophic Invasion by the Blast Fungus." *The Plant Cell* 19, no. 2 (2007): 706–24.

Kawahara, Y., Y. Oono, H. Kanamori, T. Matsumoto, T. Itoh and E. Minami. "Simultaneous RNA-Seq Analysis of a Mixed Transcriptome of Rice and Blast Fungus Interaction." *PLOS ONE* 7, no. 11 (2012): e49423. doi:10.1371/journal.pone.0049423.

Kim, S., P. Ahn and Y.-H. Lee. "Analysis of Genes Expressed During Rice-*Magnaporthe grisea* Interactions." *Molecular Plant-Microbe Interactions* 14, no. 11 (2001): 1340–6.

Koga, H., K. Dohi, O. Nakayachi and M. Mori. "A Novel Inoculation Method of *Magnaporthe grisea* for Cytological Observation of the Infection Process Using Intact Leaf Sheaths of Rice Plants." *Physiological and Molecular Plant Pathology* 64, no. 2 (2004): 67–72.

Kohli, M. M., Y. R. Mehta, E. Guzman, L. De Viedma and L. E. Cubilla. "*Pyricularia* Blast: A Threat to Wheat Cultivation." *Czech Journal of Genetics and Plant Breeding* 47 (2011): S130–4.

Kohorn, B. D. and S. L. Kohorn. "The Cell Wall-Associated Kinases, WAKs, as Pectin Receptors." *Frontiers in Plant Science* 3 (2012): 88. doi:10.3389/fpls.2012.00088.

Künstler, A., R. Bacsó, G. Gullner, Y. M. Hafez and L. Király. "Staying Alive: Is Cell Death Dispensable for Plant Disease Resistance During the Hypersensitive Response?" *Physiological and Molecular Plant Pathology* 93 (2016): 75–84.

Lannoo, N. and E. J. M. V. Damme. "Lectin Domains at the Frontiers of Plant Defense." *Frontiers in Plant Science* 5 (2014): 397. doi:10.3389/fpls.2014.00397.

Li, H., P. H. Goodwin, Q. Han, L. Huang and Z. Kang. "Microscopy and Proteomic Analysis of the Non-Host Resistance of *Oryza sativa* to the Wheat Leaf Rust Fungus, *Puccinia triticina* f. sp. *tritici*." *Plant Cell Reports* 31, no. 4 (2012): 637–50. doi:10.1007/s00299-011-1181-0.

Li, H., S.-Y. Zhou, W.-S. Zhao, S.-C. Su and Y.-L. Peng. A Novel Wall-Associated Receptor-Like Protein Kinase Gene, OsWAK1, Plays Important Roles in Rice Blast Disease Resistance." *Plant Molecular Biology* 69, no. 3 (2009): 337–46. doi:10.1007/s11103-008-9430-5.

Li, W. L., J. D. Faris, S. Muthukrishnan, D. J. Liu, P. D. Chen and B. S. Gill. "Isolation and Characterization of Novel cDNA Clones of Acidic Chitinases and β-1, 3-glucanases from Wheat Spikes Infected by *Fusarium graminearum*." *Theoretical and Applied Genetics* 102, no. 2–3 (2001): 353–62.

Liu, G., D. L. Greenshields, R. Sammynaiken, R. N. Hirji, G. Selvaraj and Y. Wei. "Targeted Alterations in Iron Homeostasis Underlie Plant Defense Responses." *Journal of Cell Science* 120, no. 4 (2007): 596–605.

Liu, G., X. Sheng, D. L. Greenshields et al. "Profiling of Wheat Class III Peroxidase Genes Derived from Powdery Mildew-Attacked Epidermis Reveals Distinct Sequence-Associated Expression Patterns." *Molecular Plant-Microbe Interactions* 18, no. 7 (2005): 730–41. doi:10.1094/MPMI-18-0730.

Liu, J., X. Wang, T. Mitchell et al. "Recent Progress and Understanding of the Molecular Mechanisms of the Rice–*Magnaporthe oryzae* Interaction." *Molecular Plant Pathology* 11, no. 3 (2010a): 419–27.

Liu, J.-J., R. Sturrock and A. K. M. Ekramoddoullah. "The Superfamily of Thaumatin-Like Proteins: Its Origin, Evolution, and Expression Towards Biological Function." *Plant Cell Reports* 29, no. 5 (2010b): 419–36.

Liu, W., J. Liu, L. Triplett, J. E. Leach and G.-L. Wang. "Novel Insights into Rice Innate Immunity Against Bacterial and Fungal Pathogens." *Annual Review of Phytopathology* 52 (2014): 213–41.

Liu, W., J. Liu, Y. Ning et al. "Recent Progress in Understanding PAMP- and Effector-Triggered Immunity Against the Rice Blast Fungus *Magnaporthe oryzae*." *Molecular Plant* 6, no. 3 (2013): 605–20.

Maciel, J. L. N., P. C. Ceresini, V. L. Castroagudin, M. Zala, G. H. J. Kema and B. A. McDonald. "Population Structure and Pathotype Diversity of the Wheat Blast Pathogen *Magnaporthe oryzae* 25 Years After its Emergence in Brazil." *Phytopathology* 104, no. 1 (2014): 95–107.

Malaker, P. K., N. C. D. Barma, T. P. Tiwari et al. "First Report of Wheat Blast Caused by *Magnaporthe oryzae* Pathotype *Triticum* in Bangladesh." *Plant Disease* 100, no. 11 (2016): 2330.

Mauch, F., C. Reimmann, E. Freydl, U. Schaffrath and R. Dudler. "Characterization of the Rice Pathogen-Related Protein Rirla and Regulation of the Corresponding Gene." *Plant Molecular Biology* 38, no. 4 (1998): 577–86. doi:10.1023/A:1006041404436.

McCarthy, D. J., Y. Chen and G. K. Smyth. "Differential Expression Analysis of Multifactor RNA-Seq Experiments with Respect to Biological Variation." *Nucleic Acids Research* 40, no. 10 (2012): 4288–97.

Medzhitov, R. "Toll-Like Receptors and Innate Immunity." *Nature Reviews Immunology* 1, no. 2 (2001): 135.

Mosquera, Gl., M. C. Giraldo, C. H. Khang, S. Coughlan and B. Valent. "Interaction Transcriptome Analysis Identifies *Magnaporthe oryzae* BAS1-4 as Biotrophy-Associated Secreted Proteins in Rice Blast Disease." *The Plant Cell* 21, no. 4 (2009): 1273–90.

Nga, N. T. T., V. T. B. Hau and Y. Tosa. "Identification of Genes for Resistance to a Digitaria Isolate of *Magnaporthe grisea* in Common Wheat Cultivars." *Genome* 52, no. 9 (2009): 801–809.

Panstruga, R. and P. N. Dodds. "Terrific Protein Traffic: The Mystery of Effector Protein Delivery by Filamentous Plant Pathogens." *Science* 324, no. 5928 (2009): 748–50.

Peng, H., Q. Zhang, Y. Li et al. "A Putative Leucine-Rich Repeat Receptor Kinase, OsBRR1, is Involved in Rice Blast Resistance." *Planta* 230, no. 2 (2009): 377–85. doi:10.1007/s00425-009-0951-1.

Picinini, E. C. and J. M. C. Fernandes. "Ocorrência da brusone (*Pyricularia oryzae*) em lavouras comerciais de trigo (*Triticum aestivum*) no estado do Rio Grande do Sul." *Fitopatologia Brasileira* 15 (1990): 83–4.

Pizolotto, C. A., J. L. N. Maciel, J. M. C. Fernandes et al. "Reação à brusone de genótipos de trigo portadores da translocação 2NS/AS de *Aegilops ventricosa*." In *XI Reunião da Comissão Brasileira de Pesquisa de Trigo e Triticale*, Fórum Nacional de Trigo, 2017, Cascavel. Resumos expandidos... Cascavel: Coodetec, 2017. 151–155.

Prabhu, A. S., M. C. Filippi and N. Castro. "Pathogenic Variation Among Isolates of *Pyricularia oryzae* Affecting Rice, Wheat, and Grasses in Brazil." *International Journal of Pest Management* 38, no. 4 (1992): 367–71.

Prestes, A. M., P. F. Arendt, J. M. C. Fernandes and P. L. Scheeren. "Resistance to *Magnaporthe grisea* Among Brazilian Wheat Genotypes." In *Wheat Production in Stressed Environments*, edited by H. T. Buck, J. E. Nisi and N. Salomón, 119–23. Dordrecht: Springer, 2007.

Ronald, P. C. and B. Beutler. "Plant and Animal Sensors of Conserved Microbial Signatures." *Science* 330, no. 6007 (2010): 1061–4.

Sasaki, K., T. Iwai, S. Hiraga et al. "Ten Rice Peroxidases Redundantly Respond to Multiple Stresses Including Infection with Rice Blast Fungus." *Plant and Cell Physiology* 45, no. 10 (2004): 1442–52. doi:10.1093/pcp/pch165.

Schweizer, P., W. Hunziker and E. Mösinger. "cDNA Cloning, *in vitro* Transcription and Partial Sequence Analysis of mRNAs from Winter Wheat (*Triticum aestivum* L.) with Induced Resistance to *Erysiphe graminis* f. sp. *tritici*." *Plant Molecular Biology* 12, no. 6 (1989): 643–54.

Serfling, A., S. E. Templer, P. Winter and F. Ordon. "Microscopic and Molecular Characterization of the Prehaustorial Resistance Against Wheat Leaf Rust (*Puccinia triticina*) in Einkorn (*Triticum monococcum*)." *Frontiers in Plant Science* 7 (2016): 1668. doi:10.3389/fpls.2016.01668.

Shimono, M., J. Yazaki, K. Nakamura et al. "cDNA Microarray Analysis of Gene Expression in Rice Plants Treated with Probenazole, a Chemical Inducer of Disease Resistance." *Journal of General Plant Pathology* 69, no. 1 (2003): 76–82. doi:10.1007/s10327-002-0010-2.

Shiu, S.-H., W. M. Karlowski, R. Pan, Y.-H. Tzeng, K. F. X. Mayer and W.-H. Li. "Comparative Analysis of the Receptor-Like Kinase Family in Arabidopsis and Rice." *The Plant Cell* 16, no. 5 (2004): 1220–34.

Silué, D., J. L. Notteghem and D. Tharreau. "Evidence of a Gene-for-Gene Relationship in the *Oryza sativa–Magnaporthe grisea* Pathosystem." *Phytopathology* 82, no. 5 (1992): 577–80.

Silva, S. R., M. C. Bassoi and J. S. S. Foloni. *Informações técnicas para trigo e triticale – safra 2017.* Brasília, DF: Embrapa, 2017.

Singh, A., S. I. Kirubakaran and N. Sakthivel. "Heterologous Expression of New Antifungal Chitinase from Wheat." *Protein Expression and Purification* 56, no. 1 (2007): 100–9.

Singh, P. and L. Zimmerli. "Lectin Receptor Kinases in Plant Innate Immunity." *Frontiers in Plant Science* 4 (2013): 124. doi:10.3389/fpls.2013.00124.

Sweigard, J. A., A. M. Carroll, S. Kang, L. Farrall, F. G. Chumley and B. Valent. "Identification, Cloning, and Characterization of PWL2, a Gene for Host Species Specificity in the Rice Blast Fungus." *The Plant Cell* 7, no. 8 (1995): 1221–33.

Tagle, A. G., I. Chuma and Y. Tosa. "Rmg7, a New Gene for Resistance to Triticum Isolates of *Pyricularia oryzae* Identified in Tetraploid Wheat." *Phytopathology* 105, no. 4 (2015): 495–9.

Takabayashi, N., Y. Tosa, H. S. Oh and S. Mayama. "A Gene-for-Gene Relationship Underlying the Species-Specific Parasitism of Avena/Triticum Isolates of *Magnaporthe grisea* on Wheat Cultivars." *Phytopathology* 92, no. 11 (2002): 1182–8.

The Gene Ontology Consortium. "Expansion of the Gene Ontology Knowledgebase and Resources." *Nucleic Acids Research* 45, Database issue (2017): D331–8. doi:10.1093/nar/gkw1108.

Thomma, B. P. H. J., T. Nürnberger and M. H. A. J. Joosten. "Of PAMPs and Effectors: The Blurred PTI–ETI Dichotomy." *The Plant Cell* 23, no. 1 (2011): 4–15.

Toledo, J. "Enfermedades." In *Guía de recomendaciones técnicas del cultivo de trigo*, edited by O. Díaz, E. Guzmán, J. Toledo, P. Franco, D. Terrazas and A. Antezana, 36–45. Santa Cruz de la Sierra, Bolivia, 2004.

Tör, M., M. T. Lotze and N. Holton. "Receptor-Mediated Signalling in Plants: Molecular Patterns and Programmes." *Journal of Experimental Botany* 60, no. 13 (2009): 3645–54. doi:10.1093/jxb/erp233.

Torres, G. A. M., H. H. Soto-Gonzáles, N. F. Webber and H. M. Scherer. "Tipos de lesiones de Piricularia en trigo." Comunicado Técnico online. 355. Passo Fundo: RS. Embrapa, 2016.

Tosa, Y. "Genetic Analysis of the Species-Specific Parasitism of Plant Pathogenic Fungi." *Journal of General Plant Pathology* 75, no. 6 (2009): 455–7. doi:10.1007/s10327-009-0193-x.

Tufan, H. A., G. R. D. McGrann, R. MacCormack and L. A. Boyd. "TaWIR1 Contributes to Post-Penetration Resistance to *Magnaporthe oryzae*, but not *Blumeria graminis* f. sp. *tritici*, in Wheat." *Molecular Plant Pathology* 13, no. 7 (2012): 653–65. doi:10.1111/j.1364-3703.2011.00775.x.

Tufan, H. A., G. R. D. McGrann, A. Magusin, J.-B. Morel, L. Miché and L. A. Boyd. "Wheat Blast: Histopathology and Transcriptome Reprogramming in Response to Adapted and Nonadapted Magnaporthe Isolates." *New Phytologist* 184, no. 2 (2009): 473–84.

Urashima, A. S., S. Igarashi and H. Kato. "Host Range, Mating Type, and Fertility of *Pyricularia grisea* from Wheat in Brazil." *Plant Disease* 77, no. 12 (1993): 1211–6.

Vaid, N., P. K. Pandey and N. Tuteja. "Genome-Wide Analysis of Lectin Receptor-Like Kinase Family from *Arabidopsis* and Rice." *Plant Molecular Biology* 80, no. 4 (2012): 365–88. doi:10.1007/s11103-012-9952-8.

Valent, B., L. Farrall and F. G. Chumley. "*Magnaporthe grisea* Genes for Pathogenicity and Virulence through a Series of Backcrosses." *Genetics* 127, no. 1 (1991): 87–101.

Van Ginkel, M., R. M. Trethowan, K. Ammar, Jiankang Wang and M. Lillemo. *Guide to Bread Wheat Breeding at CIMMYT.* Wheat Special Report No. 5 (Revised edition.). Mexico, DF: CIMMYT, 2002.

van Loon, L. C., M. Rep and C. M. J. Pieterse. "Significance of Inducible Defense-Related Proteins in Infected Plants." *Annual Review of Phytopathology* 44, no. 1 (2006): 135–62. doi:10.1146/annurev.phyto.44.070505.143425.

Vavilin, D. V. and W. F. J. Vermaas. "Regulation of the Tetrapyrrole Biosynthetic Pathway Leading to Heme and Chlorophyll in Plants and Cyanobacteria." *Physiologia Plantarum* 115, no. 1 (2002): 9–24. doi:10.1034/j.1399-3054.2002.1150102.x.

Verbon, E. H., P. L. Trapet, I. A. Stringlis, S. Kruijs, P. A. H. M. Bakker and C. M. J. Pieterse. "Iron and Immunity." *Annual Review of Phytopathology* 55, no. 1 (2017): 355–75. doi:10.1146/annurev-phyto-080516-035537.

Viedma, L. Q. and W. Paiva. *Añublo o Piricularia del Trigo. Díptico. MAG/DIA/CRIA.* Programa de Investigación de Trigo, Capitán Miranda, Itapúa – Paraguay, 2002.

Vy, T. T. P., G.-S. Hyon, N. T. T. Nga, Y. Inoue, I. Chuma and Y. Tosa. "Genetic Analysis of Host–Pathogen Incompatibility between Lolium Isolates of *Pyricularia oryzae* and Wheat." *Journal of General Plant Pathology* 80, no. 1 (2014): 59–65.

Wagner, T. A. and B. D. Kohorn. "Wall-Associated Kinases are Expressed Throughout Plant Development and are Required for Cell Expansion." *The Plant Cell* 13, no. 2 (2001): 303–18.

Wang, S., S. Asuke, T. T. P. Vy et al. "A New Resistance Gene in Combination with *Rmg8* Confers Strong Resistance Against *Triticum* Isolates of *Pyricularia oryzae* in a Common Wheat Landrace." *Phytopathology* 108, no. 11 (2018): 1299–306. doi:10.1094/PHYTO-12-17-0400-R.

Yan, L., Y. Ma, D. Liu et al. "Structural Basis for the Impact of Phosphorylation on the Activation of Plant Receptor-like Kinase BAK1." *Cell Research* 22, no. 8 (2012): 1304–8.

Yang, K., W. Rong, L. Qi, J. Li, X. Wei and Z. Zhang. "Isolation and Characterization of a Novel Wheat Cysteine-Rich Receptor-Like Kinase Gene Induced by *Rhizoctonia cerealis*." *Scientific Reports* 3 (2013): 3021. doi:10.1038/srep03021.

Yeh, Y.-H., Y.-H. Chang, P.-Y. Huang, J.-B. Huang and L. Zimmerli. "Enhanced Arabidopsis Pattern-Triggered Immunity by Overexpression of Cysteine-Rich Receptor-Like Kinases." *Frontiers in Plant Science* 6 (2015): 322. doi:10.3389/fpls.2015.00322.

Zhan, S. W., S. Mayama and Y. Tosa. "Identification of Two Genes for Resistance to Triticum Isolates of *Magnaporthe oryzae* in Wheat." *Genome* 51, no. 3 (2008): 216–21.

Zhang, S., C. Chen, L. Li et al. "Evolutionary Expansion, Gene Structure, and Expression of the Rice Wall-Associated Kinase Gene Family." *Plant Physiology* 139, no. 3 (2005): 1107–24.

Zhao, G. "Phytoferritin and its Implications for Human Health and Nutrition." *Biochimica et Biophysica Acta (BBA) – General Subjects* 1800, no. 8 (2010): 815–23. doi:10.1016/j.bbagen.2010.01.009.

6 Ecology and Epidemiology of Wheat Blast

A. E. Perelló, V. F. Consolo and I. Martinez

CONTENTS

6.1 INTRODUCTION

Magnaporthe oryzae (synonym *Pyricularia oryzae*) is a new fungal disease on wheat plants that is a serious threat and requires worldwide attention. It is considered a highly destructive pathogen given the yield failures and the significant economic losses. The disease is known as "Brusone" or "wheat blast" (WB) and was initially confined to South America but has recently emerged in Southeast Asia (Singh et al., 2016).

A typical symptom of wheat blast is the head infection, which is common and destructive in the field (Figure 6.1). The pathogen directly strikes the wheat ear and can shrivel and deform the grain (Martinez et al., 2019). Although the disease appears sporadically on wheat, other grasses are secondary hosts and for this reason

FIGURE 6.1 Spike symptoms A and leaf symptoms B.

rotations do not control it. On the other hand, the disease has an irregular frequency of outbreaks which makes it hard to understand and predict the precise conditions for disease development, or to methodically select resistant wheat lines.

The fungus is physiologically and genetically complex, so even after more than three decades, scientists do not fully understand how it interacts with wheat or which genes in wheat confer durable resistance (CIMMYT, 2016).

6.2 VARIABILITY OF *MAGNAPORTHE* SPECIES

Magnaporthe oryzae is a species complex (Couch & Kohn, 2002; Zhang et al., 2016) that causes blast disease on more than 50 species of poaceous plants, including important crops such as rice, wheat, barley, millet and oats (Urashima & Kato, 1998; Murakami et al., 2002). Several phylogenetic species (e.g., pathotypes) are proposed by cladistic analyses based on the multi-gene sequence and the host specificity and according to the original host plant from which they were isolated. Therefore, each pathotype has a restricted host range while commonly they are virulent to their original host (Tosa & Chuma, 2014). Based on seedling virulence assays with local wheat cultivars, Maciel et al. (2014) reported 14 pathotypes in the population of *M. oryzae*. Moreover, Valent et al. (2016) divided the *M. oryzae* population into *M. oryzae Oryza* pathotype (MoO), *M. oryzae Triticum* pathotype (MoT) and *M. oryzae Lolium* pathotype (MoL) causing blast disease of rice and wheat and grey leaf spot disease in annual/perennial ryegrass (*Lolium perenne*) and tall fescue (*Festuca arundinacea*), respectively. The *P. oryzae* pathotype *Triticum* is considered the subpopulation within the *P. oryzae* causal agent of wheat blast in South America

(Cruz and Valent, 2017). It has also been associated with blast disease on barley, rye, triticale and signal grass (*Urochloa* sp.) in Brazil (Verzignassi et al., 2012). Other studies have indicated the presence of a new species of *Pyricularia*, *P. graminis-tritici* (*Pygt*) causing wheat blast (Castroagudin et al., 2016).

A recent study that included 95 entire genome sequences of the wheat blast pathogen and other *Pyricularia* species reported the identification of *Pygt* as a highly diverse pathogen species distinct from *P. oryzae* that causes rice blast (species are pathogenic on wheat as well as several other hosts with a broad host range, such as signal grass, barley, oats and grasses often found as invasive species [weeds] [Castroagudin et al., 2017]). Thc fungus has high genetic and phenotypic diversity and its broad host range allows it to move between wheat and grass hosts. In this sense, the possibility of genetic recombination is thought to occur mainly on the grass hosts. The characterisation of contemporary populations of *Pygt* using genetic markers showed 198 multilocus microsatellite genotypes (MLMGs) (Castroagudin et al., 2017, Maciel et al., 2014). Moreover, other evaluations for determining the virulence of isolates on a set of wheat differentials showed 25 virulence groups from central southern Brazil (Maciel et al., 2014). Evidence from studies carried out in Brazil supports the extensive gene flow between populations of *Pygt* sampled from wheat and those sampled from at least 13 grasses (Castroagudin et al., 2017; Jank et al., 2014).

6.3 HOST RANGE

Several grasses (weeds), such as *Cenchrus echinatus*, *Eleusine indica*, *Digitaria sanguinalis*, *Brachiaria plantaginea*, *Echinocloa crusgalli*, *Pennisetum setosum*, *Hyparrhenia rufa* and *Rhynchelytrum roseum*, commonly occur in the wheat and rice fields of Brazil and are secondary hosts of *Pyricularia*, but their role in the epidemiology of wheat blast is not well understood (Prabhu et al., 1992). In Bolivia, *E. indica*, *Setaria italica*, *D. sanguinalis* and *Rottboelia exaltata* have been identified with blast symptoms (Hurtado & Toledo, 2005) as well as blast in commercially grown black oats (*Avena strigosa*) (Mehta et al., 2006). Black oats and foxtail millet (*S. italica*) are widely used in the crop rotation system of the Southern Cone region, and their susceptibility to wheat blast may well make them candidates for the primary source of infection. In artificial inoculation experiments, most of the isolates from weed species indicated that wheat is susceptible (Tredway et al., 2005; Urashima et al., 2004; Viji et al., 2001). In contrast, fungal isolates collected in nature belong to the pathotype specific to the host species from which they were collected (Couch et al., 2005; Tredway et al., 2005), indicating that cross-infection is rare in nature.

6.4 DISEASE CYCLE

The life cycle of blast previously studied on rice isolates (Wilson & Talbot, 2009) can be shared and adapted to the wheat blast disease cycle. Figure 6.2 shows the life cycle of *Magnaporthe* spp. undergoing sexual and asexual reproduction according to Dean et al. (2005).

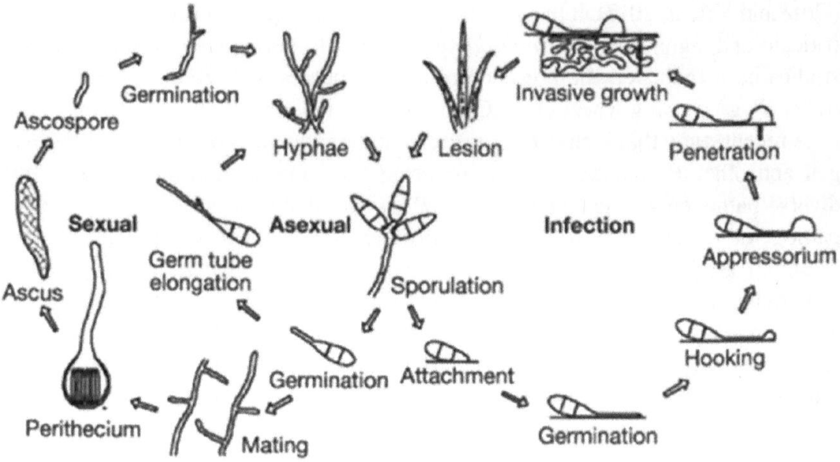

FIGURE 6.2 Life cycle of *Pyricularia oryzae*. (Adapted from Dean et al., 2005.)

To summarise, the disease cycle begins when asexual blast spores (conidia) find a host and germinate on its surface, forming an infective hypha with an appressorium at its tip. The fungus penetrates and infects the host, producing a lesion on the plant. The fungus colonises the tissues with mycelium and ends when it sporulates and disperses new airborne spores to nearby healthy plant tissues or other plants. According to Cruz et al. (2016b), although additional studies are needed to understand the significance of autoinfection in the wheat blast pathosystem. Preliminary studies in the field in Bolivia have shown that, in the absence of foliar fungicides, blast might be able to progress from older to younger leaves as the season progresses.

Cytological investigations revealed that in the case of incompatible reactions (non-host resistance) of the blast pathogen on wheat, fungal hyphae penetrated and colonised the epidermal cells and also invaded many neighbouring cells. On the other hand, in the case of compatible reactions (host resistance), if penetration did occur, the hyphae remained restricted to the first-invaded epidermal cell (Araujo et al., 2016). Additionally, unsuccessful penetration occurred with high frequency in incompatible compared to compatible interactions.

The sexual stage of the blast pathogen can be produced *in vitro* and can undergo sexual reproduction if isolates of opposite mating types are paired, but this rarely occurs in the field on any host. Ascospores are produced in asci within perithecia with long necks with the characteristics of hyaline and fusiform (spindle shaped with tapering ends) with three septae. Perithecia produce asci within 2–3 weeks when strains of opposite mating types are incubated in an oatmeal agar under light at 20°C. Mature asci are extruded through the perithecial necks and ascospores are released into a viscous liquid (Yaegashi & Udagawa, 1978). Ascospores produce appressoria for plant penetration. Sexually fertile isolates also produce small, crescent-shaped microconidia (6 μm in length and 0.7 μm in width), which are hypothesised to function as spermatia (Chuma et al., 2009; Zhang et al., 2014).

Pyricularia strains taken from grasses growing in or near infected wheat fields provided important insights into the origins, genetic structure, epidemiology and disease cycle of the wheat blast pathogen. In this way, evidence supporting the hypothesis of sexual reproduction was the formation of protoperithecia and perithecia (sexual fruiting bodies) of the fungus on autoclaved wheat stems and on senescing stems of wheat and other grasses, suggesting that sexual reproduction mainly occurs during the saprotrophic phase of the disease cycle on crop residues or within the senescent tissues of alternative hosts (Moreira et al., 2015). Moreover, *Phalaris canariensis* (canarygrass) supported the fullest development of perithecia, suggesting that it is a promising candidate for identifying the teleomorph in the field. Given that fungal strains capable of infecting both wheat and other grasses can move between different hosts, recombination may occur mainly on other grasses, giving rise to the highly diverse *Pyricularia* population. These facts confirm previous reports of cross-infection and interfertility between isolates from wheat and other poaceous hosts (Bruno & Urashima, 2001; Galbieri & Urashima, 2008; Urashima et al., 1993), the lack of differentiation among wheat and other Poaceae-adapted populations, the sharing of genotypes and virulence groups between the two host groups (Castroagudin et al., 2017) and the finding of gametic equilibrium consistent with sexual recombination in most populations, including those from other grasses.

6.4.1 Primary and Secondary Inoculum

The blast pathogen attacks the base or upper part of the rachis affecting spike formation that makes the spike partial or complete bleached or sterile resulting in shrivelled seeds or no grain (Islam et al., 2016; Prabhu et al., 1992). The affected spikelets above the infection point exhibit a bleached straw colour, which can easily be distinguished from the normal green colour of healthy spikelets (Urashima & Kato, 1994). The symptoms of wheat blast on leaves are elliptical to elongated lesions with light to dark green centres and yellow borders. However, the occurrence of disease symptoms on the leaves or stems before the heading stage is rare. In the wheat pathosystem, the development of blast signs and symptoms in all wheat aerial tissues has been reported (Igarashi, 1990). However, very little information is available concerning the *P. oryzae* inoculum produced on leaves. Despite reports (Igarashi, 1990; Murakami et al., 2000; Rios et al., 2013; Aucique Perez et al., 2014) describing blast lesions on wheat leaves, their prevalence in severely affected commercial fields is disputed (Kohli et al., 2011; Valent et al., 2013), even though it was observed that conidia production coincided with spike emergence under both greenhouse and field conditions (Cruz et al., 2015). Goulart et al. (1995) reported that the infected rachis (black point infection in rachis) passed on the pathogen to harvested seed and a significant positive correlation was found between the field incidence of wheat blast and the percentage of seed with the wheat blast pathogen across varieties. Several grasses and weeds commonly occur in wheat fields and may be secondary hosts. However, information on their role in the epidemiology of wheat blast is scant. Infected seed can play an important role in long-distance dispersal, as *PygT* is a seed-borne pathogen (Goulart & Paiva, 2000; Maciel et al., 2014). The importance of an inoculum from wheat leaves has not been determined; however, *Pyricularia* conidia produced on basal senescent leaves are considered important in the disease cycle for other plant pathosystems (Halmos, 1970).

6.4.2 Sources of Inoculum

Potential sources of primary inoculum for disease development are infected wheat seeds, crop residue or infected wheat stubble, and conidia from grass species other than wheat where the fungus overwinters (Prabhu et al., 1992; Urashima et al., 1999; Urashima & Kato, 1998; Kholi et al., 2011).

6.4.3 Seed Transmission

Goulart and Paiva (1990) have demonstrated the transmission of the fungus by wheat seed; however, little is known about the importance of seed infection in the epidemiology of the disease (Urashima et al., 1993). The presence of *P. oryzae* in the seed is directly related to germination and vigor, as this fungus can be transmitted to the progeny causing serious losses to seed producers and farmers (Teixeira & Machado, 2003). Infected seeds are one of the mechanisms of long-distance diffusion, which results in their introduction into other countries, states and regions. When infected, seeds appear small, shrivelled and with a low specific weight (Martinez et al., 2019). Consequently, most of these seeds are eliminated in the process of harvesting, which explains the low incidence of *P. oryzae* in commercial wheat seed.

Given the importance of the seed as a major source of disease transmission and the origin of the primary inoculum of infection in wheat seeds, WB needs to be considered and evaluated as a priority. In this way, the use of pathogen-free seeds is an important component of globally integrated wheat blast management worldwide.

6.4.4 Crop Residue

It is reported that the fungus *P. oryzae* survives on crop residue (CIMMYT, 2014) and plant debris (Fernandez et al., 2017). It is also reported that *P. oryzae* does not survive in stubble at the colder temperatures of Northern Indiana (Harmon & Latin, 2005), suggesting poor winter survival. Recent results by Pzolotto et al. (2016) in Brazil have shown that sporulating lesions of *Pyricularia* on wheat leaves, spikes and stems diminish over time. Moreover, according to their results, sporulation is no longer observed after 140 days of exposure. The authors concluded that it seems highly unlikely that infested wheat residues serve as a source of the wheat blast inoculum for the next wheat crop. The management of crop residues is not a key point to control the development of wheat blast. A strong emphasis should be placed on the presence of other hosts (Pizolotto et al., 2016).

6.4.5 Grass and Weeds as Hosts of *P. oryzae*

The role of grass weeds as a collateral host of *P. oryzae* in the wheat system must be considered as a possible reason for the increase in wheat blast. *P. oryzae* can infect major cereals such as rice, its best-known host, as well as wheat, barley, rye, triticale and oats. Perennial forage crops, such as signal grass, or annual crops, such as oats, which are known to be hosts of the wheat blast pathogen, are present in the landscape and thus provide a reservoir for the pathogen (Marangoni et al., 2013; Castroagudín et al., 2016). From the 21 weed species collected in the field in Paraguay, the fungus was observed on *Bromus catharticus*, *Chloris gayana*, *Cyperus diffusus*, *Digitaria horizontalis* and *Lolium multiflorum*, and pathogenicity tests conducted with isolates collected from *B. catharticus*, *D. horizontalis* and *L. multiflorum* confirmed their potential to infect wheat (Chavez & Kohli, 2015). According to Ceresini et al. (2016) in Brazil, isolates of *Pyricularia* from several poaceous hosts such as *Avena sativa*, *C. echinatus*, *Cynodon* spp., *D. sanguinalis*, *Elionurus candidus*, *Echinochloa crus-galli*, *E. indica*, *Rhynchelytrum repens* and *Urochloa* spp. were indistinguishable from *P. graminis-tritici* isolated from wheat, and populations from the two host groups had the same genetic structure. Because *Urochloa* is a widely grown pasture grass occupying more than 90 million ha in Brazil, they propose that *Urochloa* provides a major source of wheat blast inoculum and may be the preferred host for pathogen recombination.

6.4.6 Transmission and Spread

The production losses caused by wheat blast can vary from very low to almost 100%. The disease can attack all aboveground parts of the plant, but severe spike infection can be observed with very little infection on the leaves or culms. The highest losses occur when the fungus attacks the rachis at the base of the spike, thereby limiting the development of the grains and killing the spike completely.

Based on rice blast, it is assumed that conidia are the main means for the spread of the wheat blast fungus. Blast conidia appear adapted for rapid penetration of host plants but not for persistence outside the host. The dissemination of *P. oryzae* occurs primarily via airborne conidia (Ou, 1985; Urashima et al., 2007). Dispersal studies of rice blast concluded that spores could travel at least 700 m to 1 km from an inoculated source and remain viable (Ishiguro et al., 2002; Urashima et al., 2007). The long dispersal of the pathogen is through infected seeds (Urashima et al., 1999). Seed plays a limited role in the epidemiology. The infection on spikes is mainly from airborne conidia from secondary host grasses.

On the other hand, the seed-borne inoculum (Gomes et al., 2017; Goulart & Paiva 1992) has facilitated the long-distance dispersal of the pathogen and enabled it to invade other agroecosystems in South America and now Southeast Asia. Blast epidemic areas, preceding the recent emergence of *Pygt*, were reported in Bangladesh in 2016 (Islam et al., 2016).

In 1987, yield losses incurred by the three Brazilian wheat-producing states (Parana, Matto Grosso do Sul and Sao Paulo) varied between 10.5% and 53% (Goulart & Paiva, 1992). In the same year, the disease also appeared in the traditional wheat region of Rio Grande do Sul causing variable losses (Picinini & Fernandes 1990; Cunfer et al., 1993). The disease was also observed in the border region of Brazil/Paraguay in 1987. Moreover, the first epidemic in Paraguay occurred in 2002, causing production losses of more than 70% in the early seeded fields (Viedma & Morel, 2002). Most of the harvested grain did not meet commercial standards for test weight, and had to be used for animal consumption. The first severe infections of wheat blast in Bolivia were observed in the lowland Santa Cruz region in 1996 and resulted in the loss of almost 80% of production (Barea & Toledo, 1996). In 1997 (another Niño year), the disease devastated the early seeded fields, causing 100% loss, and was responsible for the severe decline in the wheat area in subsequent years. Argentina reported its first blast infections and associated losses in a summer seeded wheat experimental crop in the north-eastern state of Chaco in 2007/2008 (Alberione et al., 2008).

6.5 FAVOURABLE CONDITIONS

The literature indicates climatic conditions of frequent rainy periods, temperatures ranging from 21°C to 27°C, cloudy days and high relative humidity as most favourable for the occurrence of blast epidemics (Goulart et al., 2007; Kohli et al., 2011). The El Niño phenomenon characterised by several days of continuous rains and average temperatures between 18°C and 20°C during the flowering stage of the crop followed by sunny, hot and humid days may predispose wheat plants to blast (Saharan et al., 2016).

However, limited data are available on the environmental determinants for optimal wheat blast infection. Nowadays, it is known that the intensity of the damage caused by wheat blast is directly proportional to the level of rainfall during the heading stage of wheat. Brazil suffered one of its most serious epidemics of wheat blast in 2009 when there was a period of high precipitation during the heading stage (Maciel, 2011). Cardoso et al. (2008) developed a warning system model to predict the intensity of wheat blast based on the temperature and the duration of wheat spike wetness. The investigation was carried out at a temperature range of 10°C–35°C and a duration of spike wetness of 0–40 h. It was concluded that the wheat blast intensity could exceed 85% at 25°C and a 40 h wetness period. Alves and Fernandes (2006) revealed that the production of conidia by *Magnaporthe grisea* was favoured by high relative humidity (≥90%) and temperatures around 28°C. For the latter conditions, infection and spread to the spikes are especially effective when they are caused and perpetuated by 24 h of rain or additionally if it rains when the host is in the flowering stage. However, the severity of wheat blast varies greatly based on weather conditions, variety and the plant organ infected (Goulart et al., 2007; Urashima et al., 2009). A combination of high temperatures, excessive rain, long and frequent leaf wetness and poor fungicide efficacy has favoured the occurrence of this disease during outbreak years in Brazil (Goulart et al., 2007). It has been pointed out that the potential distribution of *P. oryzae* on wheat can be limited by specific temperature and precipitation requirements for fungal infection, and by low temperatures that might limit the overwintering survival of the fungus (Fischer, 2016).

6.6 DISEASE MONITORING IN THE FIELD

Accurate identification of the pathogen in the field is one of the measures that should be included in the prevention of possible outbreaks of the disease in areas where it has not yet established. Moreover, the certification of seeds free of the pathogen could contribute to avoiding its distribution at the international level.

These approaches require a powerful diagnostic tool with the standardisation of new techniques and methodologies for the detection assay and identification of *P. oryzae Triticum* in the field. Some of these new approaches, such as in-field whole-genome sequence (WGS) analysis for identification, are not yet practical to use. Moreover, accurate diagnostic assays are needed to distinguish the *Triticum* pathotype from all other *P. oryzae* pathotypes in a quick and safe way. The correct identification of *P. oryzae Triticum* will allow proper phytosanitary measures to be implemented to prevent further spread of the disease, as well as to track/trace outbreaks originating from foreign sources. In the study carried out by Yasuhara-Bell et al. (2018), based on Notomi et al. (2000), the PoT2 transposon and the MoT3 locus were used to develop loop-mediated isothermal amplification (LAMP) assays for the specific detection of *P. oryzae Triticum* in the laboratory and in the field. LAMP is a technology that allows the amplification of nucleic acids under isothermal conditions, enabling rapid detection in the laboratory and in the field using a variety of methods and instrumentation to perform reactions with useful results. In this study, the primers PoT2 and MoT3 demonstrated high specificity for *P. oryzae* and *P. oryzae Triticum*, respectively. Currently, given the rapid development of plant genomic technologies, the lack of access to plant phenotyping capabilities limits our ability to dissect the genetics of quantitative traits. Effective, high-throughput phenotyping platforms have recently been developed to solve this problem (Li et al., 2014), and a variety of imaging methodologies are being used to collect data for quantitative studies of complex traits related to the growth, yield and adaptation to biotic stress like diseases. These imaging techniques include visible imaging (machine vision), imaging spectroscopy (multispectral and hyperspectral remote sensing), thermal infrared imaging, fluorescence imaging, 3D imaging and tomographic imaging (magnetic resonance tomography [MRT], positron emission tomography [PET] and computed tomography [CT]). These imaging techniques and their applications in the plant phenotyping of *Pyricularia* on wheat are novel and useful tools that are just beginning to be explored.

Nowadays, the assessment of crop health status in real time could provide reliable and useful information for making effective and efficient management decisions regarding the appropriate time and method to control crop diseases. Recently, Liu et al. (2018) used the hyperspectral remote sensing methodology as a new tool for integrated pest management of the fungus on rice plants. Hyperspectral reflectance of symptomatic and asymptomatic rice leaves infected by *P. grisea* was measured in the laboratory within the 350–2,500 nm spectral region. A probabilistic neural network classifier was applied to discriminate the symptomatic rice leaves from the asymptomatic ones with the front principal component spectra (PCs). The results demonstrated that hyperspectral remote sensing can also discriminate between multiple diseases of rice leaves under laboratory conditions. Previously, Yang et al. (2012) in China, demonstrated the effectiveness of near-infrared hyperspectral imaging to detect blast rice of *Pyricularia* in Nipponbare at the seedling stage and predict the infected leaves from the healthy leaves.

Thus, there is a clear necessity to implement new tools that allow us to detect the pathogen in the field and thus help to implement the correct measures for its control.

6.7 DISEASE FORECASTING MODELLING

It is known that the development of the disease is determined by the critical period, which includes the most important factors: the temperature and the duration of the wet period. These are the factors that must be taken into account when establishing a system to predict the intensity of the disease in a certain area of cultivation. Cardoso et al. (2008) generated a model to predict the disease intensity as a function of both temperature and the duration of spike wetness and the resulting equation provided a precise description of the response of *P. oryzae* to temperatures and duration of spike wetness. This model was used to construct tables that can be used to predict the intensity of *P. oryzae* wheat blast based on the data of temperatures and the duration of wheat spike wetness observed in the field. The objective of their study was to generate basic information for developing an early warning system for wheat blast based on a climate model, for which the relationship between temperature, period of wetting of the spike and intensity of the disease was evaluated. Irrespective of the temperature, no symptoms of wheat blast were observed when the spike wetting period was less than 10 h. Moreover, the lowest blast intensity was observed at 25°C and the highest at 30°C, both with a 10 h wetting period. In excess of 85% of blast occurred at 25°C and a 40 h wetting period. The data relative to the interactions between the temperature and the duration of the spike wetting period could be used to develop a basic forecasting system based on the climatic model. The effect of the interaction between the leaf wetting period and the average temperature during the wetting period is the basis of disease forecasting systems which use the fundamental or inductive climatic model. The effect of the interactions between the temperature and the duration of the wetting period on infection occurring under controlled conditions can cause distortions when applied in the field. This is because the absolute infection frequency is difficult to predict under field conditions. One way that researchers have devised to partly avoid such problems is to divide the data on the response to infection into daily probability of infection (DPI) values. Krause and Massie (1975) and Madden et al. (1978) developed a disease forecasting system using a climatic model to develop a table of critical periods containing four arbitrary selected categories of infection efficiency, denominated daily severity values (DSV). The procedure presented in this work used the DPI values to produce a table of theoretical intensities of the disease at temperature intervals of 2°C, 3°C, 4°C and 5°C, which represent the intensity of the disease as the percentage of wheat spikes infected. These authors proposed tables that would serve as a basis to make predictions about possible wheat blast outbreaks.

Bavaresco et al. (2016) used a two-compartment model to predict wheat blast incidence. In the first compartment, inoculum potential is estimated based on the frequency of day considered favourable for sporulation. In the second compartment, spike infection risk is estimated based on the daily maximum temperature (>23°C), the temperature amplitude (>13°C) and the relative humidity (>70%). The parameterised wheat model reproduced the field observations. Finally, the wheat blast model was loosely coupled to the CSM-CROPSIM: WHEAT simulation model. The date of first spike emergence simulated by the wheat model launches the wheat blast model. A user-friendly application tool was created to run the model over multiple locations, years and sowing dates. Risk maps were produced from different years (2002–2014) and locations in the state of Parana, Brazil. In general, the disease risk maps coincided with historical observations.

More recently, Cruz et al. (2016a) generated climate-based infection models that were used to estimate the probabilities of the establishment of *P. oryzae Triticum* and outbreaks of wheat blast in the United States. Climate suitability mapping is a type of predictive mapping that can identify areas where the climatic conditions may be suitable for a pathogen to persist from one season to the next (Yang, 2003, 2006) and for disease development sufficient to cause economic loss or an epidemic greater than would normally be expected (McMichael et al., 2003; Yang, 2006). Climate suitability maps can be developed using models relying on biological information derived from controlled experiments or from the observed geographic distribution of the pest (Fowler & Takeuchi, 2012; Venette et al., 2010). In this study, specific combinations of environmental conditions for *P. oryzae Triticum* inoculum buildup prior to wheat heading, infection at the heading stage and overwintering survival of *P. oryzae Triticum* inoculum were useful in predicting the estimated likelihood of the establishment and outbreak of the disease.

Probabilities for *P. oryzae Triticum* establishment were estimated and spatially explicit accumulative risk maps developed using the formula $Pe = IBw \times IHx \times OS$, where Pe was the probability of *P. oryzae Triticum* establishing in the contiguous United States; IBw was the probability of years suitable for inoculum buildup prior to heading, with a w threshold for inoculum buildup prior to heading associated with *P. oryzae Triticum* establishment; IHx was the probability of years suitable for infection at the heading stage, with an x threshold for infection at the heading stage associated with the fungus establishment; and OS corresponded to the probability of years suitable for *P. oryzae Triticum* inoculum overwintering based on the number of days below freezing. The equation used to predict wheat blast outbreaks was $Po = IBw \times IHx \times OS$, with w and x thresholds associated with wheat blast outbreaks. According to the authors, these models allowed the significant risk of disease in some areas to be identified. With the threshold levels used, the models predicted that the climate was adequate to maintain *P. oryzae Triticum* populations in 40% of the winter wheat production areas of the United States. The threshold levels of disease outbreaks were only reached in 25% of the country. In Louisiana, Mississippi and Florida, the probability of adequate years for outbreaks was greater than 70%.

More research is needed to generate new prediction models of this disease based on the temperature and humidity ranges that are specific to each region. Such information will be useful in preventing the expansion of *P. oryzae* as well as its appearance in new areas.

6.8 EPIDEMIOLOGY AND HOST RESISTANCE TO BLAST IN ARGENTINA: A CASE STUDY

The CIDEFI (Centro de Investigaciones de Fitopatología) belonging to the FCAyF-UNLP (Facultad de Ciencias Agrarias y Forestales de la Universidad Nacional de La Plata), Buenos Aires, Argentina, has maintained a fungal collection of *P. oryzae* isolates since 2012, when the fungus was detected on wheat plants belonging to an experimental assay at IFSC, Llavallol, Lomas de Zamora, Buenos Aires, Argentina. Isolates from native woods and grasses are also included (Perelló et al., 2015). Different morphological and physiological characteristics were analysed in order to determine the divergence level among isolates. Colony colour, growth rate and sporulation on different media, temperature and light conditions are among the characteristics that have contributed most to divergence among isolates (Table 6.1).

For accurate identification, a molecular characterisation of the monosporic Argentinian isolates of *P. oryzae* from wheat was performed by end-point polymerase chain reaction (PCR) diagnostic assays using the MoT3 marker and the Pot2a primer set as a positive control (Pieck et al., 2017), confirming the presence of pathotype *Triticum* in Argentina (Figure 6.3 and Table 6.2).

TABLE 6.1
Morphocultural Characterisation of *P. oryzae* Isolates under Different Media and Light Conditions

Light conditions	Media	Isolate	Colour	Grow rate (cm/day)	Sporulation (conidia/ml)
12 h light/12 h dark	OMA	PY22	120. Pale olivaceous grey	0.66	M**
	OMA	PYPG	120. Pale olivaceous grey	0.71	L*
	OMA	PYBR	120. Pale olivaceous grey	0.79	L*
	OMA	PYBOL	120. Pale olivaceous grey	0.53	L*
	V8	PY22	121. Olivaceous grey	0.6	M**
	V8	PYPG	48. Olivaceous	0.52	M**
	V8	PYBR	69. Dark herbage green	0.53	M**
	V8	PYBOL	48. Olivaceous	0.47	M**
	RA	PY22	117. Pale mouse grey	0.56	M**
	RA	PYPG	117. Pale mouse grey	0.62	M**
	RA	PYBR	120. Pale olivaceous grey	0.6	M**
	RA	PYBOL	121. Olivaceous grey	0.54	M**
Continuous darkness	OMA	PY22	White	0.58	L*
	OMA	PYPG	White	0.61	L*
	OMA	PYBR	91. Greenish glaucous	0.61	L*
	OMA	PYBOL	122. Iron grey	0.62	L*
	V8	PY22	122. Iron grey	0.52	H***
	V8	PYPG	121. Olivaceous grey	0.58	M**
	V8	PYBR	120. Pale olivaceous grey	0.54	M**
	V8	PYBOL	133. Pale greenish grey	0.53	M**
	RA	PY22	91. Greenish glaucous	0.62	H***
	RA	PYPG	91. Greenish glaucous	0.6	L*
	RA	PYBR	91. Greenish glaucous	0.6	H***
	RA	PYBOL	91. Greenish glaucous	0.59	L*

*Low (L): 0–1.10^4 conidia/ml; **moderate (M): 1.10^4–5.10^5 conidia/ml; and ***high (H): more than 5.10^5 conidia/ml.

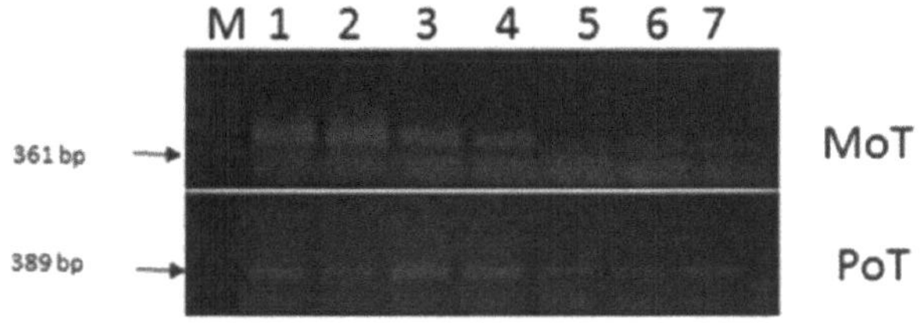

FIGURE 6.3 Agarose gel electrophoresis of PCR products from *P. oryzae Triticum* pathotype isolates 1–7 obtained from wheat samples. M: molecular marker 100 bp (Embiotec). The sizes of the products obtained after amplification of DNA samples using the set of primers MoT3F/R and Pot2a-L2/Pot2a-R/2 are indicated on the left.

TABLE 6.2
Molecular Characterisation of 50 Argentinian *P. oryzae* Isolates Tested with Two Markers

Host species	Isolated tested	Number of positive isolates for markers	
		Mot3	Pot2a
Oryza sativa	1	0	1
Triticum aestivum	50	31	50
Cynodon dactylon	1	0	1

6.8.1 Update, Priorities and Progress in Argentina

P. oryzae is a new fungal problem on wheat plants that requires global attention. The disease, known as "Brusone" or wheat blast, is recognised as a serious threat to wheat worldwide. Although currently confined to South America, *P. oryzae* is considered a highly destructive pathogen given the yield failures and the significant economic losses that it has caused in countries neighbouring Argentina (Bolivia, Paraguay and Brazil). In Brazil, the disease was first reported in the state of Parana in 1985 (Igarashi et al., 1986), later spreading to all wheat-producing regions (Anjos et al., 1996; Goulart & Paiva, 2000; Goulart et al., 1990; Igarashi, 1990; Maciel et al., 2014; Piccinini & Fernandes, 1989 and 1990). In Bolivia, the disease was first reported by Barea and Toledo (1996) and more recently in Paraguay (Viedma, 2005). Since the first detection of the disease in Chaco Province, Argentina, in 2007 (Alberione et al., 2008; Cabrera & Gutierrez, 2007), research has progressed as follows: In 2012, wheat plants were found with symptoms compatible with WB during routine monitoring of fungal diseases conducted in Llavallol, Province of Buenos Aires. In 2015, the first international report of the research period 2012–2015 at the CIDEFI was published (Perelló et al., 2015). Efforts to date aim to understand the biology and epidemiology of this new disease (Perelló, 2016); to provide knowledge about the interaction between wheat–*Pyricularia* pathotypes; and to focus attention on the susceptibility of the germplasm currently used in Argentina against the fungal infection at both the leaf and heading (spike) stages. In spite of the lack of previous information, the main actions undertaken to involve the characterisation of the fungus population from the morphocultural, molecular and biochemical point of view. Furthermore, it is necessary to improve knowledge of the aggressiveness of different pathotypes from wheat and weeds grasses and other Poaceae different from wheat and on wheat commercial cultivars currently under field cultivation. Our studies focused on the ability of the fungus as a seed-borne pathogen and its transmission to seedlings. The effect of the fungus infection on a 1000 grain weight reduction and protein content was recently analysed (Perelló et al., 2017; Martinez et al., 2019). Our research also focused on the role of the different components that participate in the cross-talk among hormone-induced signalling pathways triggered in wheat plants infected with isolates of *Pyricularia* of different aggressiveness under abiotic stress conditions such as salinity (Cabot et al., 2018).

The devastating epidemic of WB in 2016 in Asia (Bangladesh) was a warning of the global threat against other countries free of the disease (Islam et al., 2016). Although WB has not yet been detected in Europe, it is necessary to prepare for the disease. In this sense, a platform of collaboration between UNLP (National University of La Plata, Buenos Aires, Argentina) and two European Universities in Spain and Germany has been initiated. The UIB (University of Balearic Islands, Spain) project aims to understand the fungus behaviour infecting wheat plants under salinity stress and the gene defence expression. Moreover, collaboration with RWTH University (Aachen, Germany) has been initiated with the aim of deciphering the resistance of wheat against the upcoming threat of *P. oryzae* which causes blast disease. In Argentina, no new reports of field detection of the disease on wheat exist to date. However, the potential for a highly severe occurrence on barley plants under field conditions is of concern (Gutierrez & Cundom, 2015). Moreover, the disease may be masked or confused with other diseases of similar symptoms in spike such as *Fusarium* head blight (FHB) caused by *F. graminearum*, or it may go unnoticed due to the lack of a correct diagnosis. All these factors require immediate attention by the researchers and policy makers involved in wheat crop protection to prevent new outbreaks of WB.

Variability in the degree of pathogenicity was tested in artificial inoculation experiments (Perelló et al., 2017; Martinez et al., 2019). Previous documented experiments under greenhouse conditions (Perelló et al., 2017; Martinez et al., 2019) showed significant differences in lesion development on leaves (infected area [%]/total area of leaf) between isolates, and wheat varieties after foliar inoculation with *P. oryzae* at the tillering stage (Z2.1) (Zadoks, 1972). Moreover, different patterns of reaction across leaves depending on the type of strain inoculated were found, such as minute pinhead-sized spots, small brown to dark lesions with or without grey centres or typical blast lesions with elliptical shapes and grey centres. These results indicate variation in aggressiveness among 15 *P. oryzae* isolates evaluated on two varieties (Figure 6.4 and Table 6.3). Saprophytic growth and conidiation on the basal senescent leaves under greenhouse conditions were also previously reported (Cruz et al., 2015). Moreover, it was proposed that the lower canopy of wheat cultivars could play a relevant role as a source of initial inoculum for the development of wheat head blast infection and epidemics (Cruz et al., 2015). However, very little information is available in the literature and more research is needed to determine whether the inoculum

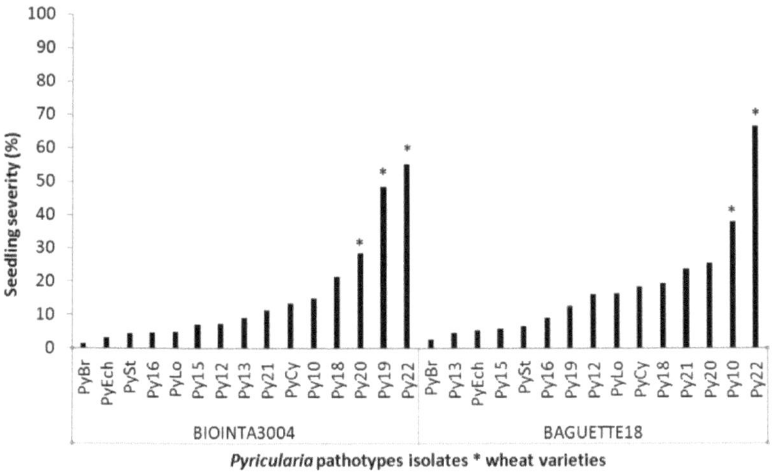

FIGURE 6.4 Severity of *Pyricularia* pathotypes isolates evaluated on varieties BioInta 3004 and Baguette 18 at seedling stage.

TABLE 6.3
Phenotypic Reaction (according to Valent et al. [1991]) on 20 Commercial Wheat Varieties Examined for Symptoms at Seedling Stage (Z1.4)

Variety/isolate	Phenotypic reaction			
	Mo33Ar	Mo28Ar	Mo12Ar	Mo6Ar
ACA315	1	3	1	1
ACA360	1	2	1	1
Baguette 11	4	4	4	4
Baguette 601	1	3	3	3
BASILIO	4	4	4	4
BIOCERES1008	1	3	2	2
BioInta 3006	1	4	4	4
BioInta 2006	1	3	3	3
BUCK APARCERO	1	2	4	0
BUCK BELLACO	2	1	1	1
ALGARROBO	0	3	3	3
CAMBIUM	0	3	1	1
KLEIN SERPIENTE	2	3	3	3
KLEIN LIEBRE	4	3	3	1
MSINTA615	4	3	4	3
MSINTABON215	1	4	4	4
SY120	0	3	4	0
SY211	1	1	1	1
TSR1066	0	2	2	2
TSR1146	0	2	4	4

originating from leaves has major relevance as a source of spike infection in severely affected commercial fields.

Concerning the impact of the disease at the spike stage, the infection patterns at anthesis (Z65 stage) (Zadocks, 1972) were evaluated by inoculating different wheat cultivars under greenhouse conditions (Perelló et al., 2017; Martinez et al., 2019). Information on the disease damage indicates highly significant differences in the susceptibility levels of the different wheat cultivars tested. Interestingly, isolates with contrasting behaviour according to the cultivar were found (Figure 6.5). These results indicate that there is an opportunity to identify large numbers of tolerant genotypes.

The seed-borne infection and transmission of *P. oryzae* on different commercial wheat cultivars currently grown in Argentina were analysed using the standard blotter test method, and differences among the wheat cultivars and isolates tested were shown. Klein Proteo germination was not affected by any of the isolates of the fungus tested; on Baguette 18, isolates PyBra and PyBol caused a significant reduction of 15% in germination. Interestingly, all the isolates tested significantly affected the germination of BioInta 3004 causing a reduction of between 50% and 66%. Symptoms observed were total or partial rot of the seeds, and necrotic symptoms with brown dots and brown or grey spot discolouration. The presence of fungal mycelium was observed and recovered under microscope. The incidence of symptoms indicated that the cultivar BioInta 3004 was the most severely infected. The incidence of 60% of grains with softening symptoms was observed on BioInta 3004 in combination with three of the five tested isolates. Moreover, seedlings that emerged from BioInta 3004 infected with isolate PyArg22 showed symptoms of the disease (necrotic spots on coleoptile and first leaf emerged). Seed transmission was evaluated on superficially disinfected and then lyophylised BioInta 3004 seeds using

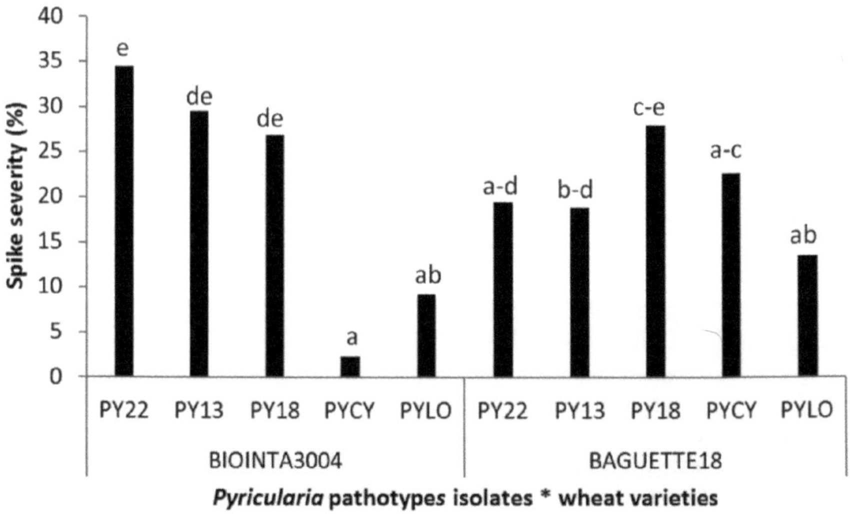

FIGURE 6.5 Severity of *Pyricularia* pathotypes isolates evaluated on varieties BioInta 3004 and Baguette 18 at heading stage.

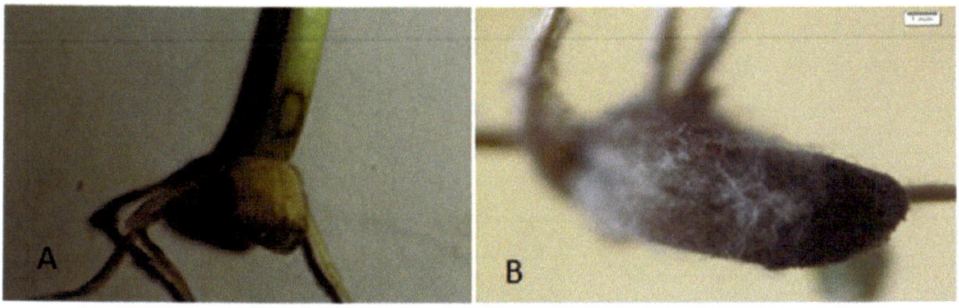

FIGURE 6.6 Transmission from seed to seedling of *P. oryzae* in wheat A, wheat seeds symptoms and the presence of mycelium B.

the PCR technique. These results were positive confirmation of the presence of *P. oryzae* in wheat seeds (Figure 6.6).

Recent results under greenhouse conditions in Argentina indicate that the fungus has the ability to grow and develop below the optimal values indicated in previous references (Com. Pers. data not shown). *In vitro* assays with two wheat isolates of *P. oryzae* pathotype *Triticum* in three different temperatures (15°C, 20°C and 25°C) and four water activity (aw) values (0.995, 0.980, 0.950 and 0.90) showed that the fungus could grow and develop in temperatures of 15°C–25°C and water activity values between 0.950 and 0.995. However, the faster growth rate was observed at the highest temperature and 0.995 wa (Larran et al., 2018).

Regarding the role of secondary hosts in the epidemiology of the disease, Perelló et al. (2015) recovered the fungus from native weed and grasses (*Stenotaphrum secundatum*, *Setaria* sp., *Eleusine* sp., *Bromus unioloides*, *L. perenne*, *Echinochloa cruz-galli*, *Avena fatua* and *Cynodon dactylon*). Most of the isolates were pathogenic on wheat plants (Figure 6.7).

It is known that the fungus affects a wide range of weeds normally found in association with cereals such as *E. indica*, *D. sanguinalis*, *Brachiaria*, *E. crus-galli*, *P. setosum*, *C. dactylon*, *L. multiflorum*, *Cyperus rotundus* and St. Augustine Grass (*S. secundatum* [Walt.] Kuntze), some of which have been identified in Argentina as secondary hosts (Martinez, 1971; Wolcan & Perelló, 1987; Pedraza et al., 2008, 2014). Nonetheless, little is known about the importance of the inoculum proceeding from secondary hosts under natural conditions.

In Argentina, results indicate that *Pyricularia* isolates from native weeds and grasses cause typical symptoms on wheat. Isolates are pathogenic for wheat plants of the cultivars BioInta 3004 and Baguette 18 under greenhouse conditions. Symptoms commonly observed were chlorotic lesions, necrotic diamond-shaped lesions, greenish necrotic lesions and elongated brown lesions, depending on the strains and cultivars analysed (Perelló et al., 2017).

6.9 PATHOGEN PHYTOTOXINS AS AGENTS OF SPECIFICITY

The fungi generally produce metabolites in their culture medium (Turner, 1971). According to their chemical structures, the metabolites can be classified into several different classes of compounds, such as secondary metabolites. Metabolites

FIGURE 6.7 Secondary host of *Pyricularia* pathotypes showing symptoms in *Echinocloa crus-galli* A, *Festuca arundinacea* B, *Digitaria sanguinalis* C and *Stenotaphrum secundatum* D.

produced by various blast disease isolates from various host plant species were reported (Nukina, 1999; Hu et al., 2011; Sato & Kozaba, 1969). *P. oryzae* produces toxins made up of different compounds e.g., pyricularin, pyriculariol and pyriculol, phytotoxic metabolites found associated with the production of a dark necrotic spot on rice leaf similar to that of blast (Ogasawara et al., 1957; Nukina et al., 1981; Rao & Suryanarayanan, 1974). Moreover, novel pyriculol-related compounds, epipyriculol, were isolated from cultures of *P. oryzae* with phytotoxic activity as well as inhibitory activity towards the spore germination of the fungus itself (Kondo et al., 1991). Similarly, Tsurushima et al. (2005) found pyrichalasin H production and the pathogenicity of *Digitaria*-specific isolates of *P. grisea*. From culture filtrates of 72 isolates of *Pyricularia* analysed via high-performance liquid chromatography (HPLC), 31 (r9 DNA type) from crabgrass (*D. sanguinalis*), 1 (r9 DNA type) from Pangolagrass (*Digitaria smutsii*) and 6 (r8 DNA type) from *D. horizontalis* produced 20–280 µg pyrichalasin H per millilitre of culture and were pathogenic on 5 *Digitaria* species. Interestingly, two isolates originally isolated from *Digitaria* plants did not produce pyrichalasin H nor cause blast lesions on *Digitaria* plants. Other reports also indicate the presence of pyriculol, pyrichalasin H and tenuazonic acid (Iwasaki et al., 1969, 1972a; Nukina, 1987). Tenuazonic acid is also produced by many isolates of *Alternaria*, but pyriculol and pyrichalasin H are specifically produced by the blast fungus. Tenuazonic acid isolated from *Oryza* isolates of the blast fungus induced typical brown necrosis on rice leaves (Umetsu et al., 1973; Iwasaki et al., 1969, 1972b). Recently, Nukina and Saito (2014) reported the production of a norvaline-analogue of tenuazonic acid by *P. oryzae*. Previously, Nukina (1988) reported the production of terrestric acid as a phytotoxic metabolite of *Pyricularia* isolates. Moreover, its phytotoxicity was demonstrated on rice plants as yellowing of the leaves and growth inhibition of the roots and shoots, indicating that terrestric acid may have had an influence on chlorophyll synthesis. The isolation of picolinic acid as a toxic substance produced by *P. oryzae* was also reported (Tamari & Kaji, 1954; Ogasawara et al., 1961). Studies on the wheat/*P. oryzae* system started at CIDEFI by analysing toxins production using the HPLC technique which showed that compounds such as picolinic acid, terrestric acid, benzoic acid and pyricurol were produced by different Argentinian isolates of *P. oryzae* from wheat in both liquid media tested (Table 6.4). On the other hand, benzoic acid, acetamide, acetic acid, furoic acid, carboxylic acid, pyrichalasin H and pyricularin were only produced in Czapek media. Other substances, e.g. nitrobenzyl, tenuazoic acid

TABLE 6.4
Secondary Metabolites Produced by Different *P. oryzae* Isolates after 14 Days Grown under Continuous Shaking at 26°C and 3 Liquid Culture Media Detected by HPLC Technique

Czapeck media	Oatmeal media	Bran wheat media
α-Teralones	Nitrobenzyl	α Picolinic acid
Terrestric acid	α-Tetralones	Pyriculol
Benzoic acid	Benzoic acid	Dihidropyriculol
Furoic acid	Tenuazoic acid	Terrestric acid
Pyricularin+	α Picolinic acid	Tenuazoic acid
Acetamide	Pyriculol	Pyriculamide+
Acetic acid	Epipyriculol	
α Picolinic acid	Dihidropyriculol	
Pyriculol	Terrestric acid	
Carboxylic acid	Pyriculariol+	
Pyrichalasin H+		

+ Compounds produced in a particular media

and pyriculariol, were produced exclusively in OM media. Interestingly, when a third media was tested, pyriculamide was produced in liquid wheat media (WM) (wheat flakes [30 g] boiled in 500 ml of distilled water for 30 min and filtered through muslin cloth and distilled water to make up a volume of 1000 ml). Currently, from the chemotaxonomical viewpoint, the blast fungi can be divided into several groups based on their metabolite production profiles. This finding indicates the biochemical plasticity of the analysed Argentinian isolates regarding secondary metabolites and the risk of some compounds, e.g. tenuazoic acid, as an important mycotoxin of the *Alternaria* species group. On the other hand, these results are new findings that contribute to further knowledge of the variability of the isolates analysed. Research on the role of the phytotoxic metabolites of *P. oryzae* wheat isolates in pathogenesis is underway even though some of them do not show host specificity.

Globally, a complete understanding of the ecological and epidemiological factors that drive wheat blast outbreaks is required. The basic aspects of wheat blast are yet to be well determined including factors that support inoculum buildup and the wide spread of the pathogen.

REFERENCES

Alberione, E., C. Bainotti, I. Cettour and J. Salines. "Evaluación de enfermedades en trigos en siembra de verano en el NEA argentino-Campaña 2007/2008." In *7 Congreso Nacional de Trigo.* Santa Rosa, La Pampa, Argentina, 2008.

Alves, K. J. P. and J. M. C. Fernandes. "Influência da temperatura e da umidade relativa do ar na esporulação de *Magnaporthe grisea* em trigo." *Fitopatologia Brasileira* 31 (2006): 579–84.

Anjos, J. R. N., D. B. Da Silva, M. J. D. Charchar and G. C. Rodrigues. "Ocorrência de brusone (*Pyricularia grisea*) em trigo e centeio na região dos cerrados do Brasil Central." *Pesquisa Agropecuária Brasileira* 31 (1996): 79–82.

Araujo, L., J. M. Soares, M. C. C. De Filippi and F. A. Rodrigues. "Cytological Aspects of Incompatible and Compatible Interactions Between Rice, Wheat and the Blast Pathogen *Pyricularia oryzae*." *Scientia Agricola* 73 (2016): 177–83.

Aucique Perez, C. E., F. Á. Rodrigues, W. R. Moreira and F. M. DaMatta. "Leaf Gas Exchange and Chlorophyll a Fluorescence in Wheat Plants Supplied with Silicon and Infected with *Pyricularia oryzae*." *Biochemistry and Cell Biology* 104 (2014): 143–9.

Barea, G. and J. Toledo. "Identificación y zonificación de piricularia o bruzone (*Pyricularia oryzae*) en el cultivo del trigo en el dpto. de Santa Cruz." In *CIAT. Informe Técnico*: Proyecto de Investigación Trigo, Santa Cruz 1996, 76–86.

Bavaresco, J. L., A. T. Lazzaretti, R. Y. Tsukahara, W. Pavan and J. M. C. Fernandes. "A Weather-Based Simulation Model for Wheat Blast Incidence." *2nd International Workshop on Wheat Blast.* April 6–9, 2016, 161. Florianopolis, SC, Brazil. 2016.

Bruno, A. C. and A. S. Urashima. "Sexual Relationship Between *Magnaporthe grisea* from Wheat and from Other Hosts." *Fitopatologia Brasileira* 26 (2001): 21–6.

Cabot, C., R. Bosch, S. Martos, C. Poschenrieder and A. Perelló. "Salinity is a Prevailing Factor for Amelioration of Wheat Blast by Biocontrol Agents." *Biological Control* 125 (2018): 81–9. doi:10.1016/j.biocontrol.2018.07.003.

Cabrera, M. G. and S. A. Gutierrez. "Primer registro de *Pyricularia oryzae* en cultivos de trigo del NE de Argentina." In *Jornada de Actualización en Enfermedades de Trigo. Lavallol*, edited by M. Astiz Gasso and M. Molina, 18–180. Buenos Aires: Instituto Fitotécnico de Santa Catalina, 2007.

Cardoso, C. A. A., E. M. Reis and E. N. Moreira. "Development of a Warning System for Wheat Blast Caused by *Pyricularia grisea*." *Summa Phytopathologica* 34 (2008): 216–21.

Castroagudín, V. L., A. Danelli, S. I. Moreira et al. "The Wheat Blast Pathogen *Pyricularia graminis-tritici* has Complex Origins and a Disease Cycle Spanning Multiple Grass Hosts." *bioRxiv* 203455, 2017. doi.10.1101/203455.

Castroagudin, V. L., S. I. Moreira, D. A. S. Pereira et al. "*Pyricularia graminis-tritici*, a New Pyricularia Species Causing Wheat Blast." *Persoonia* 37 (2016): 199–216.

Cerisini, P. C., V. L. Castrogudin, J. N. L. Maciel et al. "The Recent Emergence and Evolution of the Wheat Blast Species Complex in Brazil." *2nd International Workshop on Wheat Blast*. April 6–9, 2016, 131. Florianopolis, SC, Brazil.

Chavez, A. and M. Kohli. "Hospederos alternativos de *Magnaporthe grisea* del trigo en Paraguay." *Ciencia e investigación agraria* 17 (2015): 54–9.

Chuma, I., T. Shinogi, N. Hosogi et al. "Cytological Characteristics of Microconidia of *Magnaporthe oryzae*." *Journal of General Plant Pathology* 75 (2009): 353–8.

CIMMYT. "Annual Report. Turning Research into Impact." 2014. www.cimmyt.org/news/cimmyt-annual-report-2014-turning-research-into-impact-now-available-online/.

CIMMYT. "Annual Report." 2016. https://www.cimmyt.org/wp-content/uploads/2016/03/blastmap-web.png.

Couch, B. C. and L. M. Kohn. "A Multilocus Gene Genealogy Concordant with Host Preference Indicates Segregation of a New Species, *Magnaporthe oryzae*, from *M. grisea*." *Mycologia* 94 (2002): 683–93.

Couch, B. C., I. Fudal, M. H. Lebrun et al. "Origins of Host-Specific Populations of the Blast Pathogen Magnaporthe Oryzae in Crop Domestication with Subsequent Expansion of Pandemic Clones on Rice and Weeds of Rice." *Genetics* 170 (2005): 613–30.

Cruz, C. and B. Valent. "Wheat Blast Disease: Danger on the Move." *Tropical Plant Pathology* 42 (2017): 210–22.

Cruz, C. D., W. W. Bockus, F. M. Santana et al. "Wheat Blast: Progress in Elucidating its Ecology and Developing Strategies for Control." *2nd International Workshop on Wheat Blast*, April 6–9, 2016b, 133. Florianopolis, SC, Brazil.

Cruz, C. D., R. D. Magarey, D. N. Christie, G. A. Fowler and M. Fernandez. "Climate Suitability for *Magnaporthe oryzae Triticum* Pathotype in the United States." *Plant Disease* 100 (2016a): 1979–87.

Cruz, C. D., J. Kiyuna, W. W. Bockus, T. C. Todd, J. P. Stack and B. Valent. "*Magnaporthe oryzae* Conidia on Basal Wheat Leaves as a Potential Source of Wheat Blast Inoculum." *Plant Pathology* 64 (2015): 1491–8.

Cunfer, B. M., T. Yorinori and S. Igarashi. "Wheat Blast." In *Seed Borne Diseases and Seed Health Testing of Wheat*, edited by S. B. Mathur and B. M. Cunfer. Copenhagen: Danish Government Institute of Seed Pathology for Developing Countries, 1993.

Dean, R. A., N. J. Talbot, D. J. Ebbole et al. "The Genome Sequence of the Rice Blast Fungus *Magnaporthe grisea*." *Nature* 434 (2005): 980–6.

Fernandes, J. M. C., M. Nicolau, W. Pavan et al. "A Weather Based Model for Predicting Early Season Inoculum Build Up and Spike Infection by the What Blast Pathogen." *Tropical Plant Pathology* 42, no. 3 (2017): 230–7.

Fischer, T. "Influence of Freezing on the Survival of *Magnaporthe oryzae* and Weather Conditions that Favor Blast Epidemics in Rice." MS Thesis. Manhattan: Kansas State University, 2016.

Fowler, G. and Y. Takeuchi. "Mapping, Climate and Geographic Information for Risk Analysis." In *Plant Pest Risk Analysis: Concepts and Applications*, edited by C. Devorshak, 151–63. Oxford: CAB International, 2012.

Galbieri, R. and A. S. Urashima. "Sexual Characterization, Compatibility and Occurrence of Sexual Reproduction Among Isolates of *Pyricularia grisea* from Different Hosts." *Summa Phytopathologica* 34 (2008): 22–8.

Gomes, D. P., V. S. Rocha, O. L. Pereira and M. A. de Souza. "Damage of Wheat Blast on the Productivity and Quality of Seeds as a Function of the Initial Inoculum in the Field." *Journal of Seed Science* 39 (2017): 66–74.

Goulart, A. C. P. and F. A. Paiva. "Transmission of *Pyricularia oryzae* by Wheat *Triticum aestivum* Seeds." *Fitopatologia Brasileira* 15 (1990): 359–62.

Goulart, A. C. P. and F. A. Paiva. "Incidência da brusone (*Pyricularia oryzae*) em diferentes cultivares de trigo (*Triticum aestivum*) em condições de campo." *Fitopatologia Brasileira* 17 (1992): 321–5.

Goulart, A. C. P. and F. A. Paiva. "Perdas no rendimento de grãos de trigo causada por *Pyricularia grisea*, nos anos de 1991 e 1992, no Mato Grosso do Sul." *Summa Phytopathologica* 26 (2000): 279–82.

Goulart, A. C. P., F. A. Paiva and P. J. M. Andrade. "Relationship Between Incidence of Blast in Wheat Heads and the Presence of *Pyricularia grisea* in the Harvested Seeds." *Fitopatologia Brasileira* 20 (1995): 184–9.

Goulart, A. C. P., F. A. Paiva and A. N. Mesquita. "Occurrence of Wheat Blast (*Pyricularia oryzae*) in the State of Mato Grosso do Sul." *Fitopatologia Brasileira* 15 (1990): 112–4.

Goulart, A. C. P., P. G. Sousa and A. S. Urashima. "Damages in Wheat Caused by Infection of *Pyricularia grisea*." *Summa Phytopathologica* 33 (2007): 358–63.

Gutierrez, S. A. and M. A. Cúndom. "*Pyricularia oryzae* Affecting Barley Crops in Corrientes (Argentina) [*Pyricularia oryzae* en cultivos de cebada en Corrientes (Argentina)]." *Summa Phytopathologica* 41 (2015): 318–20.

Halmos, S. "Inoculum Sources of *Pyricularia grisea*, the Cause of Pitting Disease of Bananas." *Phytopathological Notes* 60 (1970): 183–4.

Harmon, P. F. and R. Latin. "Winter Survival of the Perennial Ryegrass Pathogen *Magnaporthe oryzae* in North Central Indiana." *Plant Disease* 89 (2005): 412–8.

Hu, Y., C. Xie, G. Yang, L. Zhang, Y. Chen and L. Tang. "Effect of Toxins from *Pyricularia oryzae* Cav on Physiological and Biochemical Indexes of Rice." *Hubei Agricultural Sciences* 50 (2011): 4–8.

Hurtado, J. and J. Toledo. "Malezas hospederas de *Piricularia*, *Pyricularia grisea*. Invierno 2004." *Informe CIAT*, 2005.

Igarashi, S. "Update on Wheat Blast (*Pyricularia oryzae*) in Brazil." In *Proceedings of the International Conference of Wheat Nontraditional Warm Areas*, edited by D. Saunders, 480–3. Mexico, D.F., 1990.

Igarashi, S., C. M. Utiamada, L. C. Igarashi, A. H. Kazuma and R. S. Lopes. "*Pyricularia* in Wheat. 1. Occurrence of *Pyricularia* sp. in Parana State." *Fitopatologia Brasileira* 11 (1986): 351–2.

Ishiguro, K., R. Harasawa and T. Kobayashi. "DNA fingerprinting Reveals How Rice Blast Disease Spreads Spatially within the Local Area." *3rd International Rice Blast Conference*, 2002, 136. Tsukuba, Japan. Abstracts.

Islam, M. T., D. Croll, P. Gladieux et al. "Emergence of Wheat Blast in Bangladesh was Caused by a South American Lineage of *Magnaporthe oryzae*." *BMC Biology* 14 (2016): 84.

Iwasaki, S., K. Okaniwa and S. Okuda. "Stereochemical Study on the Management of Rearrangements of 3p-hydroxy 4,4dialkyldo-steroids and their p-toluenesulfonates." *Tetrahedron Letters* 45 (1972a): 4601–4.

Iwasaki, S., H. Muro, S. Nozoe, S. Okuda and Z. Sato. "Isolation of 3, 4-dihydro-3, 4, 8-trihydroxy-1 (2H)-naphthalenone and Tenuazonic Acid from *Pyricularia oryzae* Cavara." *Tetrahedron Letters* 1 (1972b): 13–16.

Iwasaki, S., S. Nozoe, S. Okuda, Z. Sato and T. Kozaka. "Isolation and Structural Elucidation of a Phytotoxic Substance Produced by *Pyricularia oryzae* Cavara." *Tetrahedron Letters* 45 (1969): 3977–80.

Jank, L., S. C. Barrios, C. B. do Valle, R. M. Simeao and G. F. Alves. "The Value of Improved Pastures to Brazilian Beef Production." *Crop Pasture Science* 65 (2014): 1132–7.

Kohli, M. M., Y. R. Mehta, E. Guzman, L. de Viedma and L. E. Cubilla. "*Pyricularia* Blast – A Threat to Wheat Cultivation." *Czech Journal of Genetics and Plant Breeding* 47 (2011), 130–4.

Kondo, Y., S. Sekido, I. Yamaguchi et al. "Structures of Two Novel Pyriculol-related Compounds and Identification of Naturally Produced Epipyriculol from *Pyricularia oryzae*." *Agricultural and Biological Chemistry* 55 (1991): 2785–91.

Krause, R. A. and L. B. Massie. "Predictive Systems: Modern Approaches to Disease Control." *Annual Review of Phytopathology* 13 (1975): 31–47.

Larran, S., I. Martinez and A. Perelló. "Estudio ecofisiológico de *Pyricularia oryzae* y su biocontrol con cepas endófitas y saprófitas de *Trichoderma harzianum*." *Revista agronómica del noroeste argentino* 38 (2018): 117. Abstract.

Li, L., Q. Zhang and D. Huang. "A Review of Imaging Techniques for Plant Phenotyping." *Sensors* 14 (2014): 20078–111.

Liu, Z. Y., J. G. Qi, N. N. Wang et al. "Hyperspectral Discrimination of Foliar Biotic Damages in Rice Using Principal Component Analysis and Probabilistic Neural Network." *Precision Agriculture* 19 (2018): 973–94.

Maciel, J. L. N. "*Magnaporthe oryzae*, the Blast Pathogen: Current Status and Options for its Control." *CAB Reviews: Perspectives in Agriculture, Veterinary Science, Nutrition and Natural Resources* 6 (2011): 1–8.

Maciel, L., P. C. Ceresini, V. L. Castroagudin, M. Zala, G. H. Kema and B. A. McDonald. "Population Structure and Pathotype Diversity of the Wheat Blast Pathogen *Magnaporthe oryzae* 25 Years After its Emergence in Brazil." *Phytopathology* 104 (2014): 95–107.

Madden, L., S. P. Pennypacker and A. A. Macnab. "FAST, a Forecast System for *Alternaria solani* on Tomato." *Phytopathology* 68 (1978): 1354–8.

Marangoni, M. S., M. P. Nunes, N. Jr Fonseca and Y. R. Mehta. "Pyricularia Blast on White Oats: A New Threat to Wheat Cultivation." *Tropical Plant Pathology* 38 (2013): 198–202.

Martinez, C. A. "Mancha gris de la moha (*Setaria italica* [L.] Beauv.)." Boletin de Divulgacion Tecnica No. 7. Estacion Experimental Regional Agropecuaria, INTA, Buenos Aires, *Hemisferio Sur* 2 (1971): 106–13.

Martinez, I., A. Sanabria, F. Consolo et al. "Wheat Blast: Aggressiveness of Isolates of *Pyricularia oryzae* and Effect on Grain Quality." *Journal of King Saud University – Science* 31, no. 2 (2019): 150–7. doi:10.1016/j.jksus.2018.05.003.

McMichael, A., D. Campbell-Lendrum, C. Corvalán et al. *Climate Change and Human Health—Risk and Responses*. Geneva, Switzerland: World Health Organization, 2003.

Mehta, Y. R., M. P. Nunes and J. C. Oliveira. "Occurrência de brusone em aveia no Estado do Paraná." In *Resultados Experimentais. XXVI Reunião da Comissão Brasileira de Pesquisa de Aveia*, 55–7. Guarapuava PR: FAPA, 2006.

Moreira, S. I., P. C. Ceresini and E. Alves. "Sexual Reproduction in *Pyricularia oryzae*." *Summa Phytopathology* 41 (2015): 175–82.

Murakami, J., R. Tomita, T. Kataoka, H. Nakayashiki, Y. Tosa and S. Mayama. "Analysis of Host Species Specificity of *Magnaporthe grisea* Toward Foxtail Millet Using a Genetic Cross Between Isolates from Wheat and Foxtail Millet." *Phytopathology* 93 (2002): 42–5.

Murakami, J., Y. Tosa, T. Kataoka et al. "Analysis of Host Species Specificity of *Magnaporthe grisea* Toward Wheat Using a Genetic Cross Between Isolates from Wheat and Foxtail Millet." *Phytopathology* 90 (2000): 1060–7.

Notomi, T., H. Okayama, H. Masubuchi et al. "Loop-Mediated Isothermal Amplification of DNA." *Nucleic Acids Research* 15 (2000): 28 E63k.

Nukina, M. "Pyrichalasin H., a New Phytotoxic Metabolite Belonging to the Cytochalasans from *Pyricularia grisea* (Cooke) Saccardo." *Agricultural and Biological Chemistry* 51 (1987): 1221–4.

Nukina, M. Terrestric Acid as a Phytotoxic Metabolite from *Pyricularia oryzae* Cavara. *Agricultural and Biological Chemistry* 52, no. 9 (1988): 2357–8.

Nukina, M. "The Blast Disease Fungi and Their Metabolic Products." *Journal of Pesticide Science* 24 (1999): 293–8.

Nukina, M. and T. Saito. "Microbial Production of a Norvaline-analog of Tenuazonic Acid by *Pyricularia oryzae* Cavara." *Bioscience, Biotechnology and Biochemistry* 56 (2014): 1314–5.

Nukina, M., T. Sassa, M. Ikeda, Y. Umezawa and H. Tasaki. "Pyriculariol, a New Phytotoxic Metabolite of *Pyricularia oryzae* Cavara." *Agricultural and Biological Chemistry* 45 (1981): 2161–2.

Ogasawara, N., K. Tamari and J. Kaji. "Biochemical Studies on the Blast Disease of Rice Plants. Part 7. Pyricularin-Binding-Protein. 1. Presence of Pyricularin-Binding-Protein in the Broth Culture of *Pyricularia oryzae*." *Journal of the Agricultural Chemical Society of Japan* 31 (1957): 464–8.

Ogasawara, N., K. Tamari and M. Suga. "Biochemical Studies on the Blast Disease of Rice Plants. Part 20. Mechanism of Spore Germination of *Pyricularia oryzae*. 2. Presence of Picolinic Acid in Blast Fungal Spores and Its Discharge During the Spore Germination." *Journal of the Agricultural Chemical Society of Japan* 35 (1961): 1412–16.

Ou, S. H. *Rice Diseases*. Kew: CAB International Mycological, Institute Kew, 1985.

Pedraza, M. V., M. N. Liberman and M. N. Asselborn. "*Pyricularia grisea* en alpiste (*Phalaris canariensis* L.) en Argentina." In *1° Congreso Argentino de Fitopatología*, 109. Córdoba, 2008.

Pedraza, M. V., C. A. Liberman, E. Nuñez Bordoy and M. N. Asselborn. "Hospedantes secundarios como fuente de inóculo potencial del Quemado del Arroz en Argentina". In: Congreso Argentino de Fitopatología, 3., 2014, San Miguel de Tucumán. Anales. San Miguel de Tucumán: Asociación Argentina de Fitopatólogos, 2014. p.281. Disponível em: <http://aafitopatologos.com.ar/media/secciones/241_desc.pdf?05a317>. Acesso em: 10 dic. 2018

Perelló, A. "Update on Recent Wheat Blast Research Progress in Argentina." In *2nd International Workshop on Wheat Blast*. April 6–9, 2016, 146. Florianopolis, SC, Brazil.

Perelló, A., I. Martínez and M. Molina. "First Report of Virulence and Effects of *Magnaporthe oryzae* Isolates Causing Wheat Blast in Argentina." *Plant Disease* 99 (2015): 1177.

Perelló, A., I. Martinez, A. Sanabria, R. Altamirano and J. Sibole. "Pathogenicity of Isolates of *Magnaporthe oryzae* from Wheat and Grasses Infecting Seedlings and Mature Wheat Plants in Argentina." *Plant Pathology* 66 (2017): 1149–61.

Piccinini, E. C. and J. M. C. Fernandes. "Ocorrência da brusone (*Pyricularia oryzae*) em lavouras comerciais de trigo (*Triticum aestivum*) no estado do Rio Grande do Sul." *Fitopatologia Brasileira* 14 (1989): 125.

Piccinini, E. C. and J. M. C. Fernandez. "Ocorrência da brusone (*Pyricularia oryzae*) em lavouras comerciais de trigo (*Triticum aestivum*) no estado do Rio Grande do Sul". *Fitopatologia Brasileira* 15 (1990): 83–84

Pieck, M. L., A. Ruck, M. L. Farman et al. "Genomics-Based Marker Discovery and Diagnostic Assay Development for Wheat Blast." *Plant Disease* 101 (2017): 103–9.

Pizolotto, C. A., J. L. N. Maciel, J. M. C. Fernandes and W. Boller. "Saprophytic Survival of *Magnaporthe oryzae* in Infested Wheat Residues. In *2nd IWWB-Session 4-Ecology, Epidemiology and Management*. Book of Abstracts. April 6–9, 2016, 150. Florianopolis, SC, Brazil.

Prabhu, A. S., M. C. Filippi and N. Castro. "Pathogenic Variation Among Isolates of *Pyricularia oryzae* Infecting Rice, Wheat, and Grasses in Brazil." *Tropical Pest Management* 38 (1992): 367–71.

Rao, A. N. and S. Suryanarayanan. "Studies on the Toxins of Pyricularia 1. Detection of Pyriculol in Cultures of *Pyricularia*." *Proceedings of the Indian Academy of Sciences, Section B* 80 (1974): 31–40.

Rios, J. A., D. Debona, H. S. S. Duarte and F. A. Rodrigues. "Development and Validation of a Standard Area Diagram Set to Assess Blast Severity on Wheat Leaves." *European Journal of Plant Pathology* 136 (2013): 603–11.

Saharan, M. S., S. C. Bhardwaj, R. Chatrath, P. Sharma, A. K. Choudhary and R. K. Gupta. "Wheat Blast Disease: An Overview." *Journal of Wheat Research* 8 (2016): 1–5.

Sato, Z. and T. Kozaka. "Metabolites of *Pyricularia oryzae*: Symptoms Produced on Leaf." *Annals of the Phytopathological Society of Japan* 35 (1969): 97.

Singh, R. P., P. K. Singh and J. Rutkoski. "Disease Impact on Wheat Yield Potential and Prospects of Genetic Control." *Annual Review of Phytopathology* 54 (2016): 303–22.

Tamari, K. and J. Kaji. "On the biochemical studies of *Pyricularia oryzae* Cavara, the causative mould of the blast disease, local name 'Imochi disease', of rice plants. Part 2. The isolation of picolinic acid and piricularin, the toxic substances produced by *Pyricularia oryzae*." *Niigata Univesity Faculty Agriculture Bulletin* 5 (1954): 33–42.

Teixeira, H. and J. C. Machado. "Transmissibilidade e efeito de Acremonium strictum em sementes de milho." *Ciência e Agrotecnologia* 25 (2003): 1045–52.

Tosa, Y. and I. Chuma. "Classification and Parasitic Specialization of Blast Fungi." *Journal of General Plant Pathology* 80 (2014): 202–9.

Tredway, L. P., K. L. Stevenson and L. L. Burpee. "Genetic Structure of *Magnaporthe grisea* Populations Associated with St. Augustine Grass and Tall Fescue in Georgia." *Phytopathology* 95 (2005): 463–71.

Tsurushima, T., L. D. Don, K. Kawashima et al. "Pyrichalasin H Production and Pathogenicity of Digitaria-Specific Isolates of *Pyricularia grisea*." *Molecular Plant Pathology* 6, no. 6 (2005): 605–13. doi:10.1111/j.1364-3703.2005.00309.x.

Turner, W. B. *Fungal Metabolites*, London and New York: Academic Press, 1971.

Umetsu, N., J. Kaji and K. Tamari. "Isolation of Tenuazonic Acid from Blast-Diseased Rice Plants." *Agricultural and Biological Chemistry* 37 (1973): 451–2.

Urashima, A. S. and H. Kato. "Varietal Resistance and Chemical Control of Wheat Blast Fungus." *Summa Phytopathologica* 20 (1994): 107–12.

Urashima, A. S. and H. Kato. "Pathogenic Relationship Between Isolates of *Pyricularia grisea* of Wheat and Other Hosts at Different Host Developmental Stages." *Fitopatologia Brasileira* 23 (1998): 30–5.

Urashima, A. S., S. Igarashi and H. Kato. "Host Range, Mating Type, and Fertility of *Pyricularia grisea* from Wheat in Brazil." *Plant Disease* 77 (1993): 1211–6.

Urashima, A. S., C. Grosso, A. Stabili et al. "Effect of *Magnaporthe grisea* on Seed Germination, Yield and Quality of Wheat." In *Advances in Genetic, Genomics and Control of Rice Blast Disease*, edited by G. L. Wang and B. Valent, 267–77. Dordrecht: Springer, 2009.

Urashima, A. S., Y. Hashimoto, Y. Don et al. "Molecular Analysis of the Wheat Blast Population in Brazil with a Homolog of Retrotransposon MGR583." *Annals of the Phytopathological Society of Japan* 65 (1999): 429–36.

Urashima, A. S., N. A. Lavorent, A. C. P. Goulart and Y. R. Mehta. "Resistance Spectra of Wheat Cultivars and Virulence Diversity of *Magnaporthe grisea* Isolates in Brazil." *Fitopatologia Brasileira* 29 (2004): 511–8.

Urashima, A., S. Leite and R. Galbieri. "Eficiencia da disseminacão aérea em *Pyricularia grisea*." *Summa Phytopathologica* 33 (2007): 275–9.

Valent, B., W. Bockus, C. Cruz, J. Stack and B. Valent. "Recovery Plan for Wheat Blast Caused by *Magnaporthe oryzae* Triticum Pathotype." In *USDA National Plant Disease Recovery System*, 2013. www.ars.usda.gov/ARSUserFiles/00000000/opmp/Wheat%20Blast%20Recovery%20Plan%20Final.pdf

Valent, B., L. Farrall and F. G. Chumley. "*Magnaporthe grisea* Genes for Pathogenicity and Virulence Identified through a Series of Backcrosses." *Genetics* 127 (1991): 87–101.

Valent, B., C. D. Cruz, M. Farman et al. "Strategies for Managing Blast Disease of Wheat." In *Achieving Durable Resistance to Wheat Diseases and Pests.* 2–3 November 2016, 5. Bloomington and Minneapolis, MN: Marriott USA.

Venette, R. C., D. J. Kriticos, R. D. Magarey et al. "Pest Risk Maps for Invasive Alien Species: A Road Map for Improvement." *Bioscience* 60 (2010): 349–62.

Verzignassi, R. S., L. S. Poltronieri, R. L. Benchimol, S. K. França, E. Santos de Carvalho and C. D. Fernandes. "*Pyricularia grisea*: New Pathogen on *Brachiaria brizantha* cv. Marandu in Pará." *Summa Phytopathologica* 38 (2012): 254.

Viedma, L. Q. "Wheat Blast Occurrence in Paraguay." *Phytopathology* 95, no. 6 (2005): S152.

Viedma, L. Q. and W. Morel. *Añublo o Piricularia del Trigo.* Díptico. MAG/DIA/CRIA. Programa de Investigación de Trigo, CRIA, Capitán Miranda, Itapúa, 2002.

Viji, G., B. Wu, S. Kang, W. Uddin and D. R. Huff. "*Pyricularia grisea* Causing Gray Leaf Spot of Perennial Ryegrass Turf: Population Structure and Host Specificity." *Plant Disease* 85 (2001): 817–26.

Wilson, R. and N. Talbot. "Under Pressure: Investigating the Biology of Plant Infection by *Magnaporthe oryzae*." *Nature Reviews Microbiology* 7 (2009): 185–95.

Wolcan, S. and A. Perello. "*Magnaporthe grisea* (Hebert) Yaegashi & Udagawa Anamorfo: *Pyricularia oryzae* [Cke.] Sacc.) sobre *Stenotaphrum secundatum* (Walt.) Kuntze en la Argentina." *Revista de Agronomia del NO Argentino* 24 (1987): 171.

Yaegashi, H. and S. Udagawa. "The Taxonomical Identity of the Perfect Stage of *Pyricularia grisea* and Its Allies." *Canadian Journal of Botany* 56 (1978): 180–3.

Yang, X. B. "Risk Assessment: Concepts, Development, and Future Opportunities." Online publication. *Plant Health Progress* (2003). doi:10.1094/PHP-2003-1113-02-RV.

Yang, X. B. "Framework Development in Plant Disease Risk Assessment and its Application." *European Journal of Plant Pathology* 115 (2006): 25–34.

Yang, Y., R. Chai and Y. He. Early Detection of Rice Blast (*Pyricularia*) at Seedling Stage in Nipponbare Rice Variety using Near-Infrared Hyper-Spectral Image. *African Journal of Biotechnology* 11, no. 26 (2012): 6809–17.

Yasuhara-Bell, J., K. F. Pedley, M. Farman, B. Valent and J. P. Stack. Specific Detection of the Wheat Blast Pathogen (*Magnaporthe oryzae Triticum*) by Loop-Mediated Isothermal Amplification. *Plant Disease* 102 (2018): 2550–9.

Zadoks, J. C. "Methodology of Epidemiological Research." *Annual Review of Phytopathology* 10 (1972): 253–76.

Zhang, R. E., H. F. Zheng and Z. G. Zhang. "The *Magnaporthe grisea* Species Complex and Plant Pathogenesis." *Molecular Plant Pathology* 17 (2016): 796–804.

Zhang, H. L., Z. S. Wu, C. F. Wang, Y. Li and J. R. Xu. "Germination and Infectivity of Microconidia in the Rice Blast Fungus *Magnaporthe oryzae*." *Nature Communications* 5 (2014): 4518.

7 Effect of Blast on Wheat Physiology

Fabrício Ávila Rodrigues, Carlos Eduardo Aucique-Pérez, Daniel Debona and Jonas Alberto Rios

CONTENTS

7.1 INTRODUCTION

Blast, caused by the hemibiotroph ascomycetous *Magnaporthe oryzae* pathotype *Triticum* (MoT), also designated as *Pyricularia graminis-tritici* in some literature (Castroagudín et al., 2016), is one of the most critical diseases on wheat (*Triticum aestivum* L.) in Argentina, Bangladesh, Bolivia, Brazil and Paraguay because of its harmful effects on the quality and yield of grains (Ceresini et al., 2018; Rodrigues et al., 2017). The blast appearance in Bangladesh in 2016 called the attention of Asian governments and of many plant pathologists to search for new control methods that could be used in an integrated disease management programme and to better understand how it could affect wheat physiology (Ceresini et al., 2018; Rodrigues et al., 2017). Information regarding the diversity of the MoT population, blast control strategies and disease impact on wheat physiology and, consequently, on yield has been reported in the literature (Debona et al., 2014a; Maciel et al., 2014; Pagani et al., 2014; Rocha et al., 2014; Rodrigues et al., 2017).

Photosynthesis is one of the most important physiological processes occurring in plants, in which light is converted to chemical energy and photoassimilates are produced for use in other physiological events (Taiz & Geiger, 2006). The CO_2 fixation in plants is drastically altered by pathogen infection regardless of their lifestyles (e.g., biotrophic, hemibiotrophic and necrotrophic) (Berger et al., 2004, 2007; Bolton, 2009; Bonfig et al., 2006; Scharte et al., 2005; Zou et al., 2005). Plants exposed to different types of biotic stresses show a reduction in their photosynthetic performance mainly due to lower CO_2 intake due to stomatal limitations, drastic inhibition of the photosystems, reduction in electron transfer, more dissipation of energy by non-photochemical mechanisms, lower concentration of photosynthetic pigments, damage to the chloroplasts, changes in the source–sink ratio and down-regulation on the expression of genes involved in photosynthesis (Barón et al., 2012; Biemelt & Sonnewald, 2005; Bilgin et al., 2010). The primary metabolites derived from photosynthesis are the source of nutrients for pathogens colonising plant tissues and they are also used in different metabolic pathways on plants, in combination with secondary metabolites, to counteract the deleterious effect of pathogen infection on the plant physiology (Bolton, 2009). Moreover, pathogens manipulate the carbohydrate metabolism, culminating in a reduction in plant growth and yield (Berger et al., 2007; Bolton, 2009). The fact that the activation of host defence responses may alter the concentrations of some metabolic intermediates, plants infected by pathogens need to reprogramme their photosynthetic performance to a scenario of intense demand for nutrients. A decrease in the photosynthetic performance of infected plants is intrinsically associated with a reduction in the availability of healthy green tissue capable of absorbing light and using it to photosynthesise as the disease symptoms (chlorosis and necrosis) develop due to the action of hydrolytic enzymes, non-host-selective toxins and reactive oxygen species (ROS). Moreover, plants are capable of turning off photosynthesis and other assimilatory metabolism processes favouring respiration. At the earliest stage of pathogen infection, the transcription of genes related to photosynthesis is down-regulated to avoid oxidative damage to the photosynthetic apparatus (Berger et al., 2007; Bolton, 2009). A reduction in the photosynthetic metabolism associated with a high demand for carbohydrates as the plants respond to pathogen infection by activating their defence mechanisms often triggers changes in the carbon metabolism (Roitsch et al., 2003). Particularly for foliar pathogens, there is an increase in the activities of extracellular invertases to favour apoplastic phloem unloading and also in the activities of enzymes involved in the sink strength (Essmann et al., 2008; Hyun et al., 2011; Roitsch et al., 2003). Invertases catalyse sucrose hydrolysis to fructose and glucose, thereby increasing the hexose concentration. Fructose, glucose and hexoses are then transported into the cells by specific transporters to provide substrates for energy production, to increase the availability of nutrients for pathogen nutrition and to favour the expression of defence genes and cell death considering that sugars act as signalling molecules to regulate the expression of some genes (Koch, 1996). Additionally, an increase in hexose concentration may also cause retro-inhibition of the photosynthetic process (Bolton, 2009; Fotopoulos et al., 2003). Although the repression of photosynthesis and changes in the carbon metabolism appear to be general responses to infections by pathogens, their effects on sugar concentrations are dependent on the host–pathogen interaction (Berger et al., 2007).

This chapter seeks to provide an overview of the physiological and biochemical alterations occurring on wheat plants during the infection process of MoT.

7.2 ALTERATIONS IN THE PHOTOSYNTHETIC PERFORMANCE OF WHEAT PLANTS IN RESPONSE TO MoT INFECTION

7.2.1 Controlled Conditions

The carbon balance, which is based on the relationship among photosynthesis, photorespiration and respiration, dramatically affects plant growth and yield (Lambers et al., 1998). Infections by pathogens drastically reduce the carbon flow processes on plants by altering the radiation interception pattern and the efficiency of radiation use (Boote et al., 1980; Johnson, 1987). Alterations in the CO_2 fixation rates in infected plants have been attributed to factors such as changes in mesophyll conductance, stomatal and biochemical limitations, changes in the number or the functionality of the chloroplasts, alterations in the concentration of carbohydrates in the host tissue surrounding the infection sites and the action of a pathogen's effectors (e.g., hydrolytic enzymes, non-host-selective toxins and other molecules) that can manipulate the plant's primary metabolism (Berger et al., 2007; Debona et al., 2012; Lopes & Berger, 2001; Shtienberg, 1992). Pathogens affect the photosynthesis of a plant during the asymptomatic phase of diseases and the level of photoinhibition indicates the level of pathogen aggressiveness (Bassanezi et al., 2002; Guest & Brown, 1997).

Changes in the values of the leaf gas exchange and chlorophyll (Chl) *a* fluorescence parameters as well as in the concentrations of amino acids, carbohydrates, photosynthetic pigments and proteins on the leaves of wheat plants from cultivars differing in their basal levels of resistance to blast and infected by MoT are illustrated in Figure 7.1. Reductions in the values of the net carbon assimilation rate (A), stomatal conductance (g_s) and transpiration rate (E) were more expressive on the leaves of wheat plants from a susceptible cultivar compared to a moderately resistant cultivar upon infection by MoT (Debona et al., 2012) (Figure 7.1). The A, g_s and E values were negatively correlated with blast severity but positively correlated with the internal CO_2 concentration (C_i) (Aucique-Pérez et al., 2014). According to these studies, the anticipated lower CO_2 influx into the leaves caused by a decrease in the stomatal aperture was not a primary factor associated with lower A values on the leaves of wheat plants infected by MoT. Accordingly, reductions in A values on wheat leaves infected by MoT were related to dysfunctions at the level of the biochemical reactions involving CO_2 fixation (Debona et al., 2014a). Therefore, biochemical constraints on CO_2 fixation resulted in many dysfunctions at the photochemical level. On wheat leaves infected by MoT, reductions in the light-saturated A and a lower light saturation point were linked with a decrease in the electron transference rate which resulted in the generation of an excess of reducing power that triggered photoinhibitory events (Aucique-Pérez et al., 2017).

A semi-quantitative analysis of Chl *a* fluorescence parameters showed that during infection by MoT on wheat leaves, the maximal photosystem II quantum yield (F_v/F_m) was significantly lower by 5%, 11% and 13% at 72, 96 and 120 hours after inoculation (hai), respectively, in comparison to non-infected leaves (Aucique-Pérez et al., 2017) (Figure 7.1). During the biochemical phase of photosynthesis, the determination of ribulose-1,5-bisphosphate carboxylase/oxygenase (RuBisCO) activity allowed the inference that the initial, final and activation states of RuBisCO were

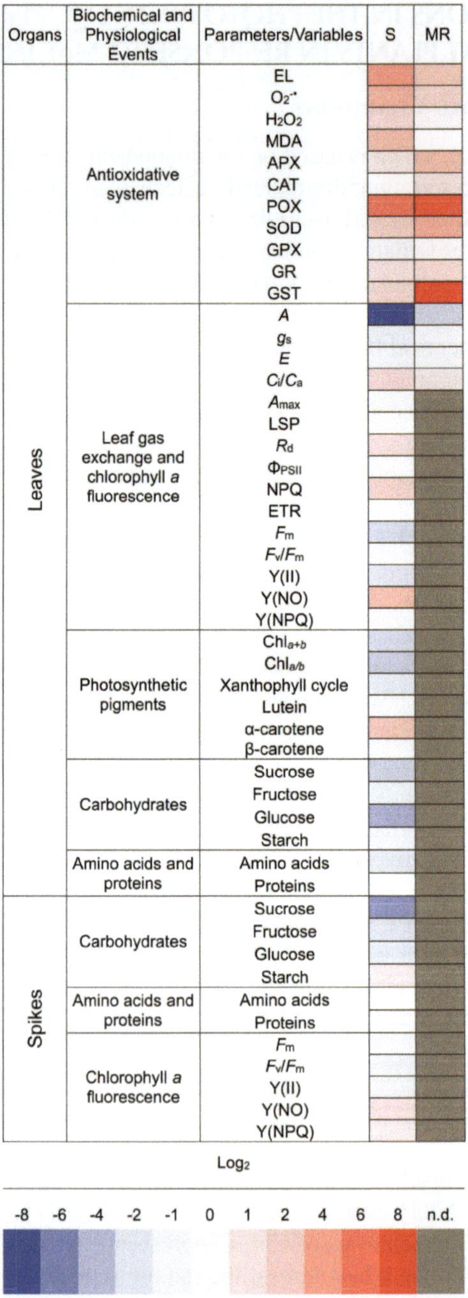

FIGURE 7.1 Biochemical and physiological changes occurring on leaves and spikes of wheat plants from susceptible (S) and moderately resistant (MR) cultivars to blast. Data from each parameter and variable were obtained after normalisation of their values from plants from S and MR cultivars inoculated with *Magnaporthe oryzae* pathotype *Triticum* (MoT) with the corresponding values from non-inoculated plants of these same cultivars. Blue and

FIGURE 7.1 (CONTINUED)
red regions indicate the levels of decrease and increase, respectively, for the parameters and variables evaluated. Data used in this analysis were compiled from several articles (Aucique-Pérez et al., 2014, 2017; Debona et al., 2012, 2014a; Rios et al., 2017). Abbreviations: A = net CO_2 assimilation rate; A_{max} = maximal net carbon assimilation rate; APX = ascorbate peroxidase; CAT = catalase; C_i/C_a = internal to ambient CO_2 concentration; Chl_{a+b} = total chlorophylls (Chl); $Chl_{a/b}$ = Chl_a/Chl_b ratio; E = transpiration rate; EL = electrolyte leakage; ETR = electron transport rate; F_m = maximal fluorescence; F_v/F_m = variable-to-maximum Chl a fluorescence ratio; GPX = glutathione peroxidase; GR = glutathione reductase; g_s = stomatal conductance to water vapour; GST = glutathione-S-transferase; H_2O_2 = hydrogen peroxide; LSP = light saturation point; MDA = malondialdehyde; n.d. = non-determined; NPQ = quenching non-photochemical; $O^{2-\bullet}$ = anion superoxide; POX = peroxidase; R_d = respiration; SOD = superoxide dismutase; Y(II) = photochemical yield; Y(NO) = quantum yield of non-regulated energy dissipation; Y(NPQ) = yield for dissipation by down-regulation; and Φ_{PSII} = photochemical yield of photosystem II.

drastically reduced and associated with an increase in blast severity. These findings confirm that biochemical limitations related to the reduced activity of RuBisCO rather than diffusive impairments were associated with the decreases in A values during MoT infection on wheat leaves. Considering that photosynthesis was drastically constrained upon MoT infection, there was a lower availability of photoassimilates that could be exported towards the sink (Rios et al., 2017).

Perturbations in the CO_2 assimilation on leaves infected by pathogens are coupled with changes in the carbohydrate metabolism as a strategy to manipulate the host physiology to efficiently take up the nutrients on the colonised tissues (Major et al., 2010). The activity of cell wall acid invertases on the flag leaves of wheat plants infected by MoT was significantly higher when compared to non-infected flag leaves, suggesting its participation in depressing sucrose pools by increasing hexoses production (Rios et al., 2017) (Figure 7.2). Moreover, there was an increase in the hexoses-to-sucrose ratio as blast symptoms developed, mainly due to an induction in cell wall invertases (Rios et al., 2017). On the other hand, this finding corroborates with the decrease in A values on leaves infected by pathogens of different lifestyles, and the accumulation of carbohydrates disrupts the translocation of assimilates to the sink (Berger et al., 2007). The results reported for the wheat–MoT interaction at the physiological level (Aucique-Pérez, 2016; Debona et al., 2014a; Rios et al., 2017) were similar to those reported for different types of grasses before the incubation period of blast (Parker et al., 2009).

Plants infected by pathogens show a drastic reduction in their concentration of photosynthetic pigments, especially Chl a and Chl b (Shtienberg, 1992). In wheat leaves infected by MoT, there were significant reductions in the concentration of photosynthetic pigments (Chl a, Chl b and carotenoids) due to the action of hydrolytic enzymes and non-host-selective toxins released by the fungus on the colonised leaf tissues (Rios et al., 2014) (Figure 7.1). The concentrations of total Chl, carotenoids (involved in the xanthophyll's cycle [antheraxanthin + violaxanthin + zeaxanthin]), β-carotene and lutein as well as the Chl_a/Chl_b ratio decreased in wheat leaves infected by MoT (Aucique-Pérez et al., 2017) (Figure 7.1). Therefore, a lower carotenoid concentration may have limited the protective capacity of the wheat plants by

FIGURE 7.2 Alterations in the activities of ADP-glucose pyrophosphorylase (AGPase), sucrose-phosphate synthase (SPS), sucrose synthase (SuSy) and invertases on the expression of α- and β-amylases genes as well as on the concentrations of soluble sugars and starch during the infection process of *Magnaporthe oryzae* pathotype *Triticum* (MoT) in the flag leaves and spikes of wheat plants. Down and up arrows indicate increase and decrease, respectively.

favouring the formation of ROS in the light phase of photosynthesis causing membrane photodamage in addition to harming the Cyt $b_{6/f}$ complex by photobleaching promoted by singlet oxygen (Zhang et al., 1999). These responses may cause damage to the chloroplasts which affects the photosynthesis of the infected plants (Barón et al., 2012; Debona et al., 2012).

7.2.2 Field Conditions

Blast greatly affects the quality and yield of wheat grains especially when MoT infection occurs in the flag leaves and spikes (Rios et al., 2017). The evaluation of healthy leaf area duration (HAD), healthy leaf area absorption (HAA), radiation intercepted (RI) and radiation use efficiency (RUE) is used to quantify the impact of diseases on yield losses (Johnson, 1990; Long et al., 2006; Moreira et al., 2015; Polanco et al., 2014; Waggoner & Berger, 1987). Rios et al. (2017) investigated the physiological performance and yield of wheat plants from cultivars Guamirim and BR-18 (susceptible and moderately resistant to blast, respectively, at the adult plant growth stage) artificially inoculated with MoT and sprayed or non-sprayed with fungicide (13% epoxiconazole + 5% pyraclostrobin) in field conditions. The use of fungicide and host resistance resulted in a higher mean grain yield and lowered the effect of blast on the physiology of the plants (Rios et al., 2017). The relationship between blast severity and grain yield was significantly negative and the greatest rates of yield reduction per unit increase in the disease were linked to the blast occurring at the early grain-filling stage in the spikes. The authors concluded that blast was likely to have a greater effect on grain yield when fungal infection occurred during the early stages of grain development than during the mid to late grain-filling stage. This finding is consistent with the results from a study carried out by Goulart et al. (2007) who observed that 63% yield loss caused by blast on wheat occurred during the early stages of grain development compared to 46% yield loss when fungal infection occurred during the latter stages of grain development. To minimise losses, blast control should be implemented on plants at their early grain-filling stage. In wheat fields where blast epidemics occur, the mean grain yields were approximately two-fold higher on plants sprayed with fungicide twice (the first and second spray at early anthesis and milk stage, respectively) in comparison to non-sprayed plants (Pagani et al., 2014). The effect of blast on grain yield was associated with the health of the leaf tissue and the photosynthetic performance of the wheat plants. A cultivar + fungicide combination with the lowest mean blast intensities on leaves and spikes generally resulted in the highest values for the healthy leaf area index (HLAI), HAA, RI by the healthy leaf area (HRI), F_v/F_m, effective PSII quantum yield Y(II) and the lowest quantum yield of regulated energy dissipation (Y(NPQ)) values. Moreover, the significant and positive relationship between HAA or HRI and yield at the early stages of grain development emphasises the importance of the healthy tissues on plants and the best photosynthetic performance of the wheat plants, particularly during the grain-filling stage. Photosynthesis on flag leaves and spikes during their initial development are important sources of photoassimilates for grain formation and, consequently, for high yield (Sanchez-Bragado et al., 2014; Tuncel & Okita, 2013). High values for leaf health (HAD, HLAI, HAA and HRI) and photosynthetic performance (based on F_v/F_m, maximum fluorescence [F_m], Y(II) and Y(NPQ) values) in both leaves and spikes were obtained using fungicide spray and a resistant cultivar in comparison to non-sprayed plants from a susceptible cultivar. These results regarding the health of wheat plants indicate the importance of maintaining non-infected leaves, particularly during the early grain-filling stage and provide other approaches for the estimation of yield and yield loss on wheat fields exposed to blast epidemics. Based on the data from non-sprayed wheat plants of the susceptible cultivar, blast affected the amount of light intercepted and absorbed by the leaves, both of which are of pivotal importance for photosynthesis and, consequently, to achieve great yield. Non-sprayed plants showed the highest values for blast severity and correspondingly the lowest measures of leaf health (HAD, HLAI,

HAA and HRI) and photosynthetic performance (based on F_v/F_m, Y(II) and Y(NPQ) values). Conversely, plants from the moderately resistant cultivar sprayed with fungicide showed the healthiest leaves and greater interception and absorption of light. For blast on leaves and spikes, the highest rates of yield reduction were associated with disease development within the first 10–14 days after anthesis and also with their best photosynthetic performance based on Chl *a* fluorescence parameters F_m, F_v/F_m, Y(II), quantum yield of non-regulated energy dissipation in PSII (Y(NO)) and Y(NPQ) (Figure 7.3). Blast severity on spikes during the early grain-filling stage had a greater effect on yield compared to blast severity on leaves possibly due to a reduction in the translocation of photoassimilates to the developing grains due to necrosis on the rachis (Rios et al., 2017). It is plausible to assume that blast on spikes gives a better estimate of its impact on yield. From a practical standpoint, the evaluation of blast on leaves and spikes of plants at anthesis and early grain-filling stages may be more useful when it aims to develop yield loss models and establish management decision thresholds. The combination of a preventative fungicide spray at anthesis for plants from a resistant cultivar can increase the photosynthetic performance of leaves and spikes, thereby achieving great yield.

7.3 METABOLISM OF CARBOHYDRATES ON WHEAT PLANTS INFECTED BY MoT

Lower yield caused by blast on the rachis of wheat plants is associated with a reduced translocation of photoassimilates to the spikes which results in shrivelled, deformed and lightweight grains (Goulart et al., 2007). The lower flow of photoassimilates in the rachis may be associated with two adverse events that may or may not occur simultaneously, caused by MoT infection in spikes: (i) the colonisation of rachis tissue by MoT blocks the nutrients translocation due to the degradation of the vascular tissue and (ii) the physiological and biochemical impairments of the photosynthetic performance of the plants and the source–sink relationship on infected rachis tissue (Figures 7.1–7.3). Carbohydrates resulting from the photosynthesis of non-stressed plants are used in sugar-importing sink organs. The yield of wheat grains is a dynamic process mainly associated with the capacity of the source leaves to transport photoassimilates to the grains and their transformation into starch (Tuncel & Okita, 2013). The CO_2 fixed in the leaves during the day is converted into sucrose in the cytosol and transported to sinks (Zeeman et al., 2007). Part of the CO_2 fixed is stored in the chloroplasts as transient starch that is broken down at night to support nocturnal leaf metabolism and sucrose exportation (Zeeman et al., 2007). On plants infected by pathogens, the homeostasis in the source–sink interface is affected via impairments on photosynthesis and on the concentration of carbohydrates and their partitioning within the infected tissues (Abood & Lösel, 2003, Gamm et al., 2011). The activities of key enzymes associated with carbon metabolism such as adenosine diphosphate (ADP)-glucose pyrophosphorylase (AGPase) and sucrose-phosphate synthase (SPS) (enzymes that catalyse the rate-limiting step of starch and sucrose biosynthesis, respectively) as well as invertases (ubiquitous enzymes in the sink that promote the cleavage of sucrose into hexoses) are negatively impacted on plant tissues infected by pathogens (Berger et al., 2007; Fotopoulos et al., 2003; Gamm et al., 2011). High invertase activity at the pathogens' infection sites is associated with a reduction in the export of photoassimilates and the unloading of sucrose into surrounding cells (Berger et al., 2004, 2007; Fotopoulos et al., 2003; Roitsch et al., 2003). The effect of blast on the source–sink interface in wheat plants was associated with alterations in the production of photoassimilates and their partitioning during the grain-filling stage (Rios et al., 2017). The concentrations of amino acids, carbohydrates and proteins decreased on the leaves and spikes of wheat plants of a susceptible cultivar to blast (Figure 7.1). The photosynthetic performance of the flag leaves and spikes of wheat plants infected by MoT was down-regulated through impairments on photosynthesis (alterations in the values of Chl *a* fluorescence parameters coupled with a reduction in or degradation of photosynthetic pigments) (Figures 7.1 and 7.3) and on CO_2 fixation (decreases in RuBisCO activity and in its activation state). The infection of the flag leaves and spikes of wheat plants by MoT affected the production of carbohydrates and their partitioning during the grain-filling stage (Rios et al., 2017) (Figure 7.1).

The photosynthesis of the flag leaves and spikes of wheat plants infected by MoT progressively decreased as blast developed. The F_v/F_m values obtained from both the flag leaves and spikes infected by MoT were below 0.80 suggesting the occurrence of photoinhibition on photosynthesis (Figure 7.3). Additionally, a reduction in the values

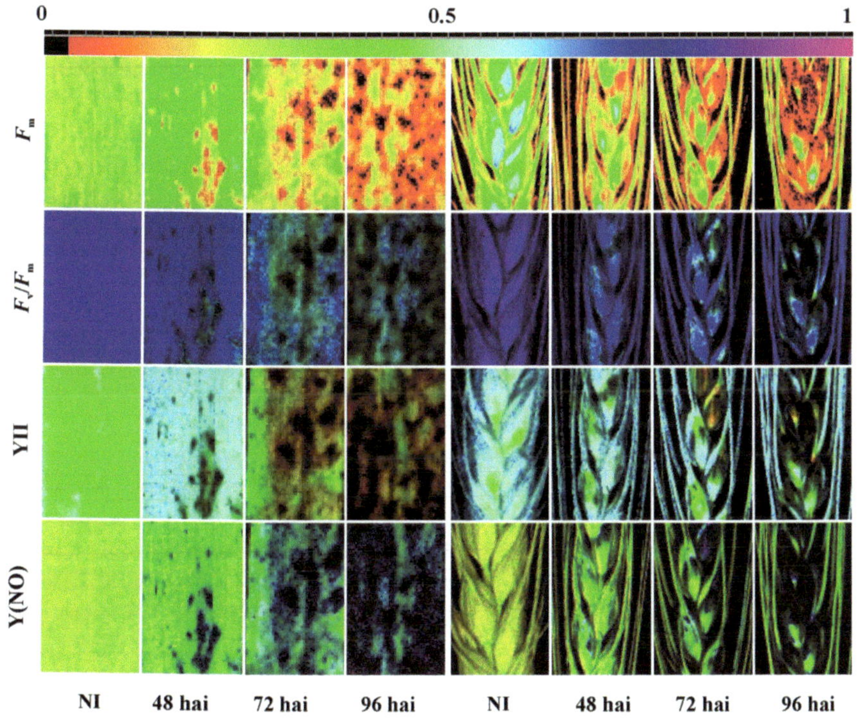

FIGURE 7.3 Parameters of chlorophyll *a* fluorescence: maximal fluorescence (F_m), maximal photosystem II quantum yield (F_v/F_m), effective PSII quantum yield (Y(II)) and quantum yield of non-regulated energy dissipation (Y(NO)) determined in the flag leaves and spikes of wheat plants from cultivar Guamirim non-inoculated (NI) or at 48, 72 and 96 hours after inoculation (hai) with *Magnaporthe oryzae* pathotype *Triticum* (MoT). Numbers from 0 to 1 on the scale bar represent the pixel value for each colour.

of Y(II) associated with a lower concentration of pigments indicated impairments in the apparent electron transport activity in the infected flag leaves and spikes of wheat plants (Figures 7.1 and 7.3). Indeed, these changes were linked with decreases and increases in Y(NPQ) and Y(NO) values, respectively (Figures 7.1 and 7.3). These alterations likely reflected the inability of the wheat plants to regulate their mechanisms of photoprotection resulting in photooxidative damage to the infected tissues. A reduction in adenosine triphosphate production can be due to photochemical dysfunctions and lower reducing power of CO_2 assimilation (Rolfe & Scholes, 2010). Nonetheless, it was expected that CO_2 fixation would also be impaired given the remarkable decrease in RuBisCO activity in the infected flag leaves of wheat plants (Rios et al., 2017). Considering the leaf gas exchange parameters examined, a reduction in photosynthesis on the leaves of wheat plants infected by MoT was governed chiefly by biochemical constraints linked to lower RuBisCO activity (Aucique-Pérez et al., 2014; Debona et al., 2014a). In any case, biochemical limitations, likely related to reduced RuBisCO activity rather than diffusive limitations, were of great importance to the lower *A* values obtained for plants infected by MoT, causing the lower availability of photoassimilates to be exported towards the sink organs.

Lower sucrose concentration, the major form of sugar transported in plants, was found in infected wheat plants suggesting changes in the metabolism at the source–sink interface (Rios et al., 2017) (Figure 7.2). A reduction in sucrose concentration was associated with less availability of substrates (triose-phosphate due to lower CO_2 fixation) and lower SPS activity (Rios et al., 2017) (Figure 7.2). Apart from lower sucrose synthesis, high invertase activity associated with the up-regulation of invertase gene expression in infected flag leaves may have contributed to depress the sucrose pools by increasing sucrose breakdown into hexoses (Figure 7.2). Notably, fructose and glucose concentrations were lower as blast in the flag leaves developed, but to a lesser extent compared to sucrose. Consequently, the hexoses-to-sucrose ratio increased as blast developed and was linked with induction in cell wall invertases. The reduction in photosynthesis coupled with a high hexoses-to-sucrose ratio suggests a sink status in the infected tissues of wheat plants. An increase in sucrose uptake and hexoses accumulation may be an indication of additional sinks in plants favouring the nutrition of pathogens (Fotopoulos et al., 2003).

Lower sucrose synthase activity associated with lower sucrose concentration in grains caused an impairment of the sinking ability of infected spikes of wheat plants (Rios et al., 2017) (Figure 7.2). The sucrose unloaded into the sink organs can be used either by cytosolic invertases or by SuSy. The SuSy uses sucrose to produce fructose and uridine diphosphate glucose (UDP-Glc), which is the first step in the utilisation of sucrose in grains (Winter & Huber, 2000). According to Li et al. (2013), higher SuSy activity may increase the sucrose flow towards the sink tissues, which may ultimately increase starch synthesis. A lower sucrose concentration in grains from spikes infected by MoT was associated with lower SuSy and invertase activities (Rios et al., 2017) (Figure 7.2). The cell wall invertase gene was induced in infected spikes of wheat plants and was accompanied by a decrease in invertase activity at the advanced stages of MoT infection (Rios et al., 2017) (Figure 7.2). It is possible that post-transcriptional regulation of invertases occurred. Regardless of these considerations, a decrease in the sucrose concentration in grains from infected spikes could be associated with a lower sucrose supply to grains rather than with an increase of sucrose use considering that the activities of sucrose-metabolising enzymes (e.g., SuSy and invertases) decreased in grains from infected spikes (Rios et al., 2017) (Figure 7.2). An increase in the starch concentration on the flag leaves of wheat plants at the advanced stage of MoT infection indicated that a reduced amount of substrate for starch synthesis was produced (reduced photosynthesis) associated with a lower expression of the ADP-glucose pyrophosphorylase gene and, consequently, with lower AGPase activity (Rios et al., 2017) (Figure 7.1). Starch is found in plant tissues infected by pathogens (Kim et al., 2009). It is possible to infer that starch accumulation is linked with a down-regulation of the starch degradation pathway considering that expressions of the α- and β-amylases genes involved in the initial attack on the starch granules were repressed on the flag leaves (Scheidig et al., 2002). In contrast to what was discovered on the flag leaves of wheat plants, starch pools decreased on infected grains (Rios et al., 2017). This finding was associated with lower sucrose transport from the photosynthetic organs coupled with lower expression of the *AGPase* gene and reduced AGPase activity (Rios et al., 2017) (Figure 7.3). Moreover, high expressions of the α- and β-amylases genes occurred at early stages of the MoT infection resulting in an increase in the breakdown of starch on infected grains (Rios et al., 2017) (Figure 7.2). These alterations should greatly impair the endosperm filling process resulting in small and shrunken grains on spikes of wheat plants infected by MoT. The carbon and nitrogen metabolisms were compromised by MoT infection considering the decreases in the concentrations of amino acids and proteins on infected flag leaves and grains (Rios et al., 2017) (Figure 7.1). According to Dupont and Altenbach (2003), the accumulation of protein in wheat grains is an essential factor associated with their development and with flour quality. A reduction in the quality of wheat grains will affect the concentration of proteins. The deleterious effects of blast on the quality and yield of grains are likely to be associated with photosynthetic dysfunctions at the photochemical and biochemical levels, impairment of sucrose synthesis and export from the flag leaves to spikes due to reduced SPS activity and also to lower starch and protein concentrations in grains (Rios et al., 2017). This results not only from the lower expression of the *AGPase* gene and reduced AGPase activity associated with a higher starch breakdown by α- and β-amylases at the early stages of MoT infection, but also from a decrease in the sucrose supply (Rios et al., 2017). The effect of MoT on the source–sink interface in wheat plants is mainly due to alterations in the production of photoassimilates and their partitioning during the grain-filling process (Rios et al., 2017).

7.4 ALTERATIONS ON THE ANTIOXIDANT SYSTEM OF WHEAT PLANTS INFECTED BY MoT

The ROS, including hydrogen peroxide (H_2O_2), hydroxyl (OH^-), singlet oxygen ($^1O_2$) and superoxide (O_2^-), are continuously produced by plants as byproducts of aerobic metabolic processes such as photosynthesis and respiration (Mittler, 2017). The ROS are generated in different cellular compartments including chloroplasts, mitochondria and peroxisomes. In turn, plants can maintain a steady state by continually scavenging these molecules using the redox machinery of the cells under normal physiological conditions (Apel & Hirt, 2004; Mittler, 2017). Unbalanced production and the removal of ROS cause oxidative stress resulting in damage to essential macromolecules such as DNA, lipids and proteins (Apel & Hirt, 2004). The equilibrium between the production and scavenging of ROS may be perturbed by many abiotic and biotic types of stresses (Apel & Hirt, 2004; Mittler, 2002, 2017). Infection by pathogens can rapidly raise the intracellular levels of ROS, with O_2^- the earliest ROS generated (Bispo et al., 2014; Fortunato et al., 2015; Noctor et al., 2002). The response of the antioxidant metabolism (concentration of ROS and enzymes activities) of plants from some wheat cultivars in response to MoT is intrinsically dependent on their level of basal resistance to blast (Figure 7.1) (Debona et al., 2012, 2014a). Debona et al. (2012) determined the concentration of ROS on wheat plants from two cultivars differing in their levels of basal resistance to blast. According to these authors, the comparison between non-inoculated and inoculated plants showed that MoT infection significantly increased O_2^- concentration by 153% and 72% at 96 hai for susceptible and moderately resistant cultivars, respectively (Figure 7.1). Similarly, H_2O_2 concentration in plants from a susceptible cultivar significantly increased by 59% at 96 hai (Figure 7.1). The peroxidation of unsaturated lipids of cell membranes represents the most prominent marker for oxidative stress in plants; therefore, the extent of the cell damage stemming from an increase in ROS production as a plant's response against pathogen attack can be estimated by quantifying the products of such membrane's lipid peroxidation (Mandal et al., 2008). Malondialdehyde (MDA), a thiobarbituric acid reactive substance, has been used to measure lipid peroxidation. The MDA concentration in wheat plants of a susceptible cultivar infected by MoT more than doubled at 96 hai in comparison to non-infected plants; nevertheless, lower increases in the MDA concentration occurred in plants from a moderately resistant cultivar (Debona et al., 2012) (Figure 7.1). In another study, although the MDA concentration in wheat plants infected by MoT increased after 24 hai, a later (96 hai) and lower increase in the MDA concentration was obtained for plants supplied with silicon (Si) which was correlated with the Si-afforded wheat resistance to leaf blast (Debona et al., 2014a). Considering that the permeability of the cell membrane on wheat leaves is lost as a result of MoT infection, increases in the extravasation of the cell's contents should be expected. In support of this hypothesis, electrolyte leakage (EL) on the leaves of wheat plants of a susceptible and a moderately resistant cultivar infected by MoT increased by 124% and 87% at 72 hai and by 323% and 188% at 96 hai, respectively (Debona et al., 2012) (Figure 7.1). Based on these findings, it can be concluded that MoT infection increases the production of O_2^- and H_2O_2 and, consequently, the oxidative stress as evidenced by the high MDA concentration and EL values, whereas wheat plants from resistant cultivars to blast are more prone to counteract the MoT-triggered oxidative stress.

Considering the cytotoxic effect of the ROS, plants have evolved complex arrays of a non-enzymatic and enzymatic system for their scavenging. Ascorbate and glutathione are essential metabolites involved in the balance of ROS metabolism (Noctor & Foyer, 1998), whereas major ROS-scavenging enzymes include ascorbate peroxidase (APX), catalase (CAT), glutathione peroxidase (GPX), glutathione-*S*-transferase (GST), glutathione reductase (GR) and superoxide dismutase (SOD) being the GR indirectly involved in the glutathione recycling (Mittler, 2002, 2017;

Mittler et al., 2004, Noctor et al., 2012). The SOD stands out as the first line of defence in preventing the oxidative stress from catalysing the dismutation of O_2^- to H_2O_2 and O_2 (Giannopolitis & Ries, 1977). Consistently with O_2^- concentration, increased SOD activity occurred in leaves of wheat plants infected by MoT in comparison to non-infected leaves (Debona et al., 2012, 2014a). Interestingly, the leaves of wheat plants from a moderately resistant cultivar showed higher increases in SOD activity which resulted in a lower concentration of O_2^- in comparison to plants from a susceptible cultivar (Debona et al., 2012) (Figure 7.1). Since high levels of O_2^- can benefit some pathogens due to ROS-induced nutrients leakage from their host cells, a higher SOD activity could be a strategy used for plants from a resistant cultivar to restrict their colonisation by removing the excess O_2^- (Debona et al., 2012). By contrast, the increase in wheat resistance to blast potentiated by Si was associated with lower increases in SOD activity indicating that, in some cases, increases in this enzyme activity may reflect higher rates of O_2^- generation stemming from the massive fungal colonisation on leaf tissues (Debona et al., 2014b). The balance between SOD and the different H_2O_2-scavenging enzymes in plant cells is considered crucial in preventing the oxidative stress caused by the uncoupled removal of O_2^- and H_2O_2 (Mittler, 2002). Considering that H_2O_2 increased on wheat leaves infected by MoT, the activities of enzymes involved in its scavenging were most likely to be induced. The CAT is a critical enzyme involved in the removal of excess H_2O_2 in peroxisomes and is indispensable for ROS detoxification during plant stress (Mittler, 2002). The CAT activity in the wheat leaves of plants from a susceptible cultivar was found to be MoT inhibited and resulted in a higher H_2O_2 concentration (Debona et al., 2012) (Figure 7.1). The decrease in CAT activity is thought to be a result of enhanced proteolysis caused by peroxisomal endopeptidases generated during the oxidative stress (Palma et al., 2002) and could benefit the colonisation of MoT on wheat leaves. By contrast, MoT failed to depress CAT activity and increase H_2O_2 concentration on the leaves of wheat plants from a moderately resistant cultivar, confirming its importance in constraining H_2O_2 generation during fungal infection (Debona et al., 2012) (Figure 7.1). The ascorbate–glutathione cycle comprises a series of redox reactions involving four enzymes named APX, dehydroascorbate reductase, GR and monodehydroascorbate reductase (Noctor & Foyer, 1998). The APX has a high affinity (μM range) for H_2O_2 and uses the ascorbate as a specific electron donor to scavenge it in the chloroplasts, mitochondria and peroxisomes (Mittler, 2002). The levels of *GR* gene transcripts as well as GR activity increased in several plant species when infected by pathogens of different lifestyles (El-Zahabi et al., 1995; Harrach et al., 2008; Sarowar et al., 2005). Accordingly, increases in APX activity were noticed on leaves of wheat plants infected by MoT (Debona et al., 2012, 2014a); however, at 96 hai such increases were significantly higher on the leaves of plants from a moderately resistant cultivar in comparison to plants from a susceptible one (Debona et al., 2012) (Figure 7.1).

Glutathione plays multiple functions in plants mainly as an antioxidant agent. Glutathione acts in thiol–disulphide interactions in which reduced glutathione (GSH) is continuously oxidised to a disulphide form (GSSG) that is recycled to GSH by NADPH-dependent GR (Noctor et al., 2012). Two lines of evidence have shed light on the importance of GR in mitigating MoT-triggered oxidative stress in wheat leaves. Firstly, investigation of GR in two wheat cultivars showed that its activity increased only for plants from a moderately resistant cultivar infected by MoT (Debona et al., 2012) (Figure 7.1). Secondly, GR appeared to be essential for wheat resistance to leaf blast owing to Si supply since a significant and negative correlation occurred between GR activity and blast severity on leaves (Debona et al., 2014a). Therefore, increases in both APX and GR activities in MoT-infected wheat leaves were linked with the role of the ascorbate–glutathione cycle in counteracting fungal-induced oxidative stress.

The GST comprises a large and diverse group of enzymes that conjugate GSH to a variety of cytotoxic, electrophilic and usually hydrophobic substrates; also, GST reduces H_2O_2 and phospholipid hydroperoxides (Dixon et al., 2010). The GST activity increased in wheat leaves infected by MoT (Debona et al., 2012). Interestingly, among the antioxidant enzymes, GST was the one whose activity was more correlated with wheat resistance to leaf blast; at 96 hai, GST activity increased by 61% for plants of a susceptible cultivar infected by MoT, whereas for plants from a moderately resistant cultivar such increase was 14 times higher (Debona et al., 2012) (Figure 7.1). Based on these findings, GST seems to play a significant role in mitigating MoT-induced oxidative stress on wheat leaves.

The GPX is another enzyme that uses GSH to reduce H_2O_2 and lipid hydroperoxides (Quan et al., 2008) and is involved in lignin biosynthesis, the degradation of indole-3-acetic acid and host resistance against diseases (Asada, 1992). The GPX activity increased on wheat leaves infected with MoT but only for plants from a susceptible cultivar (Debona et al., 2012) (Figure 7.1). The Si potentiation of wheat resistance to blast was linked to lower GPX activity (Debona et al., 2014b).

Plants are endowed with different classes of peroxidase (POX) with multiple functions. Remarkably, besides its involvement in lignin biosynthesis, POX plays a crucial role in H_2O_2 removal (Moural et al., 2017). Debona et al. (2012, 2014a) reported an increase in POX activity in the wheat leaves of plants of a moderately resistant cultivar infected by MoT than for a susceptible one (Figure 7.1). Interestingly, biochemical analysis of wheat plants infected by MoT revealed that POX was one of the most active antioxidant enzymes against MoT infection with increases of 437% and 678% at 96 hai for a susceptible and a moderately resistant cultivar, respectively (Debona et al., 2012) (Figure 7.1). This finding suggests that POX played a pivotal role in the scavenging of H_2O_2 generated during MoT infection on wheat leaves.

REFERENCES

Abood, J. K. and D. M. Losel. "Changes in Carbohydrate Composition of Cucumber Leaves During the Development of Powdery Mildew Infection." *Plant Pathology* 52 (2003): 256–65.

Apel, K. and H. Hirt. "Reactive Oxygen Species: Metabolism, Oxidative Stress, and Signal Transduction." *Annual Review of Plant Biology* 55 (2004): 373–99.

Asada, K. "Ascorbate Peroxidase: A Hydrogen Peroxide-Scavenging Enzyme in Plants." *Physiologia Plantarum* 85 (1992): 235–41.

Aucique-Pérez, C. E. "Wheat Resistance to Blast Using a Non-Host Selective Toxin and Host Metabolic Reprogramming Through a Successful Infection by *Pyricularia oryzae*." PhD Thesis, 128, Universidade Federal de Viçosa (UFV), Viçosa, Minas Gerais, Brazil, 2016.

Aucique Pérez, C. E., F. A. Rodrigues, W. R. Moreira and F. M. DaMatta. "Leaf Gas Exchange and Chlorophyll a Fluorescence in Wheat Plants Supplied with Silicon and Infected with *Pyricularia oryzae*." *Phytopathology* 104 (2014): 143–9.

Aucique-Pérez, C. E., P. E. M. Silva, W. R. Moreiras, F. M. DaMatta and F. A. Rodrigues. "Photosynthesis Impairments and Excitation Energy Dissipation on Wheat Plants Supplied with Silicon and Infected with *Pyricularia oryzae*." *Plant Physiology and Biochemistry* 121 (2017): 196–205.

Barón, M., J. Flexas and E. Delucia. "Photosynthesis Responses to Biotic Stress." In *Terrestrial Photosynthesis in a Changing Environment. A Molecular, Physiological and Ecological Approach*, edited by J. Flexas, F. Loreto and H. Medrano, 331–50. Cambridge, UK: Cambridge University Press, 2012.

Bassanezi, R. B., L. Amorim, A. Bergamin Filho and R. D. Berger. "Gas Exchange and Emission of Chlorophyll Fluorescence During the Monocycle of Rust, Angular Leaf Spot and Anthracnose on Bean Leaves as a Function of Their Trophic Characteristics." *Journal of Phytopathology* 150 (2002): 37–47.

Berger, S., A. K. Sinha and T. Roitsch. "Plant Physiology Meets Phytopathology: Plant Primary Metabolism and Plant–Pathogen Interactions." *Journal of Experimental Botany* 58 (2007): 4019–26.

Berger, S., M. Papadopoulos, U. Schreiber, W. Kaiser and T. Roitsch. "Complex Regulation of Gene Expression, Photosynthesis and Sugar Levels by Pathogen Infection in Tomato." *Physiologia Plantarum* 122 (2004): 419–28.

Biemelt, S. and U. Sonnewald. "Plant–Microbe Interactions to Probe Regulation of Plant Carbon Metabolism." *Journal of Plant Physiology* 163 (2005): 307–18.

Bilgin, D. D., J. A. Zavala, J. Zhu, S. J. Cluogh, D. Ort and E. H. DeLucia. "Biotic Stress Globally Down Regulates Photosynthesis Genes." *Plant Cell and Environment* 33 (2010): 1597–613.

Bispo, W. M. S., L. Araújo, M. B. Bermúdez-Cardona, I. S. Cacique, F. M. DaMatta and F. A. Rodrigues. "*Ceratocystis fimbriata*-induced Changes in the Antioxidative System of Mango Cultivars." *Plant Pathology* 64 (2014): 627–37.

Bolton, M. D. "Primary Metabolism and Plant Defense: Fuel for the Fire." *Molecular Plant-Microbe Interactions* 22 (2009): 487–97.

Bonfig, K. B., U. Schreiber, A. Gabler, T. Roitsch and S. Berger. "Infection with Virulent and Avirulent *P. syringae* Strains Differentially Affects Photosynthesis and Sink Metabolism in Arabidopsis Leaves." *Planta* 225 (2006): 1–12. doi:10.1007/s00425-006-0303-3.

Boote, K. J., J. W. Jones, G. H. Smerage, C. S. Barfield and R. D. Berger. "Photosynthesis of Peanut Canopies as Affected by Leaf Spot and Artificial Defoliation." *Agronomy Journal* 72 (1980): 247–52.

Castroagudín, V. L., S. I. Moreira, D. A. S. Pereira et al. "*Pyricularia graminis-tritici*, the New *Pyricularia* Species Causing Wheat Blast." *Persoonia* 37 (2016): 199–216.

Ceresini, P. C., V. L. Castroagudín, F. A. Rodrigues et al. "Wheat Blast: Past, Present, and Future." *Annual Review of Phytopathology* 56 (2018): 427–56.

Debona, D., F. A. Rodrigues, J. A. Rios and K. J. T. Nascimento. "Biochemical Changes in the Leaves of Wheat Plants Infected with *Pyricularia oryzae*." *Phytopathology* 102 (2012): 1121–9.

Debona, D., F. A. Rodrigues, J. A. Rios, S. C. V. Martins, L. F. Pereira and F. M. DaMatta. "Limitations to Photosynthesis in Leaves of Wheat Plants Infected by *Pyricularia oryzae*." *Phytopathology* 104 (2014a): 34–9.

Debona, D., F. A. Rodrigues, J. A. Rios, K. J. T. Nascimento and L. C. Silva. "The Effect of Silicon on Antioxidant Metabolism of Wheat Leaves Infected by *Pyricularia oryzae*." *Plant Pathology* 63 (2014b): 581–9.

Dixon, D. P., M. Skipsey and R. Edwards. "Roles for Glutathione Transferases in Plant Secondary Metabolism." *Phytochemistry* 71 (2010): 338–50.

Dupont, F. M. and S. B. Altenbach. "Molecular and Biochemical Impacts of Environmental Factors on Wheat Grain Development and Protein Synthesis." *Journal of Cereal Science* 38 (2003): 133–46.

El-Zahabi, H. M., G. Gullner and Z. Király. "Effects of Powdery Mildew Infection of Barley on the Ascorbate-Glutathione Cycle and other Antioxidants in Different Host–Pathogen Interactions." *Phytopathology* 85 (1995): 1225–30.

Essmann, J., I. Schmitz-Thom, H. Schön, S. Sonnewald, E. Weis and J. Scharte. "RNA Interference-Mediated Repression of Cell Wall Invertase Impairs Defense in Source Leaves of Tobacco." *Plant Physiology* 147 (2008): 1288–99.

Fortunato, A. A., D. Debona, A. M. A. Bernardeli and F. A. Rodrigues. "Changes in the Antioxidant System in Soybean Leaves Infected by *Corynespora cassiicola*." *Phytopathology* 105 (2015): 1050–8.

Fotopoulos, V., M. J. Gilbert, J. K. Pittman, A. C. Marvier, A. J. Buchanan and N. Sauer. "The Monosaccharide Transporter Gene, *AtSTP4*, and the Cell-Wall Invertase, *Atβfruct1*, are Induced in Arabidopsis During Infection with the Fungal Biotroph *Erysiphe cichoracearum*." *Plant Physiology* 132 (2003): 821–9.

Gamm, M., M. C. Héloir, R. Bligny et al. "Changes in Carbohydrate Metabolism in *Plasmopara viticola*-infected Grapevine Leaves." *Molecular Plant-Microbe Interactions* 24 (2011): 1061–73.

Giannopolitis, C. N. and S. K. Ries. "Superoxide Dismutases I. Occurrence in Higher Plants." *Plant Physiology* 59 (1977): 309–14.

Goulart, A. C. P., P. G. Sousa and A. S. Urashima. "Danos em trigo causados pela infecção de *Pyricularia grisea*." *Summa Phytopathologica* 33 (2007): 358–63.

Guest, D. and J. Brown. "Infection Processes." In *Plant Pathogens and Plant Diseases*, edited by J. F. Brown and H. J. Olsen, 245–62. Armidale, NSW, Australia: University of New England, 1997.

Harrach, B. D., J. Fodor, M. Pogány, J. Preuss and B. Barna. "Antioxidant, Ethylene and Membrane Leakage Responses to Powdery Mildew Infection of Near-isogenic Barley Lines with Various Types of Resistance." *European Journal of Plant Pathology* 121 (2008): 21–33.

Hyun, T. K., S. H. Eom, Y. G. Rim and J. S. Kim. "Alteration of the Expression and Activation of Tomato Invertases During *Botrytis cinerea* Infection." *Plant Omics* 4 (2011): 413–7.

Johnson, K. B. "Defoliation, Disease, and Growth: A Reply." *Phytopathology* 77 (1987): 1495–7.

Johnson, K. B. "Assessing Multiple Pest Population and their Effects on Crop Yield." In *Crop Loss Assessment in Rice*, edited by P. S. Teng, 203–13. Manila, The Philippines: International Rice Research Institute, 1990.

Kim, J. S., U. S. Sagaram, J. K. Burns, J. L. Li and N. Wang. "Response of Sweet Orange (*Citrus sinensis*) to *Candidatus Liberibacter asiaticus* Infection: Microscopy and Microarray Analyses." *Phytopathology* 99 (2009): 50–7.

Koch, K. E. "Carbohydrate-modulated Gene Expression in Plants." *Annual Review of Plant Physiology and Plant Molecular Biology* 47 (1996): 509–40.

Lambers, H., F. S. Chapin and T. L. Pons. *Plant Physiological Ecology*, 2nd ed. New York: Springer, 1998.

Li, J., E. Baroja-Fernandez, A. Bahaji et al. "Enhancing Sucrose Synthase Activity Results in Increased Levels of Starch and ADP-glucose in Maize (*Zea mays* L.) Seed Endosperms." *Plant and Cell Physiology* 54 (2013): 282–94.

Long, S. P., X. G. Zhu, S. L. Naidu and D. R. Ort. "Can Improvement in Photosynthesis Increase Crop Yields?" *Plant, Cell & Environment* 29 (2006): 315–30.

Lopes, D. B. and R. D. Berger. "The Effects of Rust and Anthracnose on the Photosynthetic Competence of Diseased Bean Leaves." *Phytopathology* 91 (2001): 212–20.

Maciel, J. L., P. C. Ceresini, V. L. Castroagudin, M. Zala, G. H. Kema and B. A. McDonald. "Population Structure and Pathotype Diversity of the Wheat Blast Pathogen *Magnaporthe oryzae* 25 Years After its Emergence in Brazil." *Phytopathology* 104 (2014): 95–107.

Major, I. T., M. C. Nicole, S. Duplessis and A. Séguin. "Photosynthetic and Respiratory Changes in Leaves of Poplar Elicited by Rust Infection." *Photosynthesis Research* 104 (2010): 41–8.

Mandal, S., A. Mitra and N. Mallick. "Biochemical Characterization of Oxidative Burst During Interaction Between *Solanum lycopersicum* and *Fusarium oxysporum* f. sp. *lycopersici*." *Physiological and Molecular Plant Pathology* 72 (2008): 56–61.

Mittler, R. "Oxidative Stress, Antioxidants and Stress Tolerance." *Trends in Plant Science* 7 (2002): 405–10.

Mittler, R. "ROS are Good." *Trends in Plant Science* 22 (2017): 11–19.

Mittler, R., S. Vanderauwera, M. Gollery and F. Van Breusegem. "Reactive Oxygen Gene Network of Plants." *Trends in Plant Science* 9 (2004): 490–8.

Moreira, E. N., F. X. R. Vale, P. A. Paul, F. A. Rodrigues and W. C. Jesus Júnior. "Temporal Dynamics of Soybean Rust Associated with Leaf Area Index in Soybean Cultivars of Different Maturity Groups." *Plant Disease* 99 (2015): 1216–26.

Moural, T. W., K. M. Lewis, C. Barnaba et al. "Characterization of Class III Peroxidases from Switchgrass." *Plant Physiology* 173 (2017): 417–33.

Noctor, G. and C. H. Foyer. "Ascorbate and Glutathione: Keeping Active Oxygen Under Control." *Annual Review of Plant Physiology and Plant Molecular Biology* 49 (1998): 249–79.

Noctor, G., A. Mhamdi, S. Chaouch et al. "Glutathione in Plants: An Integrated Overview." *Plant, Cell and Environment* 35 (2012): 454–84.

Noctor, G., S. D. Veljovic-Jovanovic, S. Driscoll, L. Novitskaya and C. H. Foyer. "Drought and Oxidative Load in the Leaves of C_3 Plants: A Predominant Role for Photorespiration?" *Annals of Botany* 89 Spec No (2002): 841–50.

Pagani, A. P. S., A. C. Dianese and A. C. Café-Filho. "Management of Wheat Blast with Synthetic Fungicides, Partial Resistance and Silicate and Phosphite Minerals." *Phytoparasitica* 42 (2014): 609–17.

Palma, J. M., L. M. Sandalio, F. J. Corpas, M. C. Romero-Puertas, I. McCarthy and L. A. del Río. "Plant Proteases, Protein Degradation, and Oxidative Stress: Role of Peroxisomes." *Plant Physiology and Biochemistry* 40 (2002): 521–30.

Parker, D., M. Beckmann, H. Zubair et al. "Metabolomic Analysis Reveals a Common Pattern of Metabolic Re-Programming During Invasion of Three Host Plant Species by *Magnaporthe grisea*." *The Plant Journal* 59 (2009): 723–37.

Polanco, L. R., F. A. Rodrigues, K. J. T. Nascimento et al. "Photosynthetic Gas Exchange and Antioxidative System in Common Bean Plants Infected by *Colletotrichum lindemuthianum* and Supplied with Silicon." *Tropical Plant Pathology* 39 (2014): 35–42.

Quan, L. J., B. Zhang, W. W. Shi and H. Y. Li. "Hydrogen Peroxide in Plants: A Versatile Molecule of the Reactive Oxygen Species Network." *Journal of Integrative Plant Biology* 50 (2008): 2–18.

Rios, J. A., F. A. Rodrigues, D. Debona and L. C. Silva. "Photosynthetic Gas Exchange in Leaves of Wheat Plants Supplied with Silicon and Infected with *Pyricularia oryzae*." *Acta Physiologiae Plantarum* 36 (2014): 371–9.

Rios, J. A., V. S. Rios, C. E. Aucique-Pérez et al. "Alteration of Photosynthetic Performance and Source–Sink Relationships in Wheat Plants Infected by *Pyricularia oryzae*." *Plant Pathology* 66 (2017): 1496–507.

Rocha, J. R. A. S. C., A. J. B. Pimentel, G. Ribeiro and M. A. Souza. "Eficiência de fungicidas no controle da brusone em trigo." *Summa Phytopathologica* 40 (2014): 347–52.

Rodrigues, F. A., J. A. Rios, D. Debona and C. E. Aucique-Pérez. "*Pyricularia oryzae*–Wheat Interaction: Physiological Changes and Disease Management Using Mineral Nutrition and Fungicides." *Tropical Plant Pathology* 42 (2017): 223–9.

Roitsch, T., M. E. Balibrea, M. Hofmann, R. Proels and A. K. Sinha. "Extracellular Invertase: Key Metabolic Enzyme and PR Protein." *Journal of Experimental Botany* 54 (2003): 513–24.

Rolfe. S. A. and J. D. Scholes. "Chlorophyll Fluorescence Imaging of Plant–Pathogen Interactions." *Protoplasma* 247 (2010): 163–75.

Sanchez-Bragado, R., A. Elazab, B. Zhou et al. "Contribution of the Ear and the Fag Leaf to Grain Filling in Durum Wheat Inferred from the Carbon Isotope Signature: Genotypic and Growing Conditions Effects." *Journal of Integrative Plant Biology* 56 (2014): 444–54.

Sarowar, S., E. N. Kim, Y. J. Kim et al. "Overexpression of a Pepper Ascorbate Peroxidase-like 1 Gene in Tobacco Plants Enhances Tolerance to Oxidative Stress and Pathogens." *Plant Science* 169 (2005): 55–63.

Scharte, J., H. Schon and E. Weiss. "Photosynthesis and Carbohydrate Metabolism in Tobacco Leaves During an Incompatible Interaction with *Phytophthora Nicotianae*." *Plant Cell Environment* 28 (2005): 1421–35. doi:10.1111/j.1365-3040.2005.01380.x.

Scheidig, A., A. Fröhlich, S. Schulze, J. R. Lloyd and J. Kossmann. "Down Regulation of a Chloroplast-Targeted Beta-Amylase Leads to a Starch-Excess Phenotype in Leaves." *Plant Journal* 30 (2002): 581–91.

Shtienberg, D. "Effects of Foliar Diseases on Gas Exchange: A Comparative Study." *Phytopathology* 82 (1992): 760–5.

Taiz, L. and E. Geiger. *Plant Physiology*. 4th ed. 700. Sunderland, MA: Sinauer Associates, Inc., 2006.

Tuncel, A. and T. W. Okita. "Improving Starch Yield in Cereals by Over-Expression of ADP Glucose Pyrophosphorylase: Expectations and Unanticipated Outcomes." *Plant Science* 211 (2013): 52–60.

Waggoner, P. E. and R. D. Berger. "Defoliation, Disease, and Growth." *Phytopathology* 77 (1987): 393–8.

Winter, H. and S. C. Huber. "Regulation of Sucrose Metabolism in Higher Plants: Localization and Regulation of Activity of Key Enzymes." *Critical Reviews in Biochemistry and Molecular Biology* 35 (2000): 253–89.

Zeeman, S. C., S. M. Smith and A. M. Smith. "The Diurnal Metabolism of Leaf Starch." *Biochemical Journal* 401 (2007): 13–28.

Zhang, H., D. Huang and W. A. Cramer. "Stoichiometrically Bound β-carotene in the Cytochrome b6f Complex of Oxygenic Photosynthesis Protects Against Oxygen Damage." *Journal of Biological Chemistry* 274 (1999): 1581–7.

Zou, J., S. Rodriguez-Zas, M. Aldea et al. "Expression Profiling Soybean Response to *Pseudomonas syringae* Reveals New Defense Related Genes and Rapid HR-specific Down Regulation of Photosynthesis." *Molecular Plant-Microbe Interaction* 18 (2005): 1161–74.

8 Screening Wheat Germplasm for Blast Resistance
Methods and Protocols

Alice Chavez and Man Mohan Kohli

CONTENTS

8.1 INTRODUCTION

Wheat blast disease caused by the fungus *Magnaporthe oryzae (*Catt.) (synonym *Pyricularia oryzae*) pathotype *Triticum* (MoT) has been endemic in the tropical regions of South America for over 30 years, and was recently observed in Bangladesh (Kohli et al., 2011; Malaker et al., 2016). When it was first identified in Londrina, Parana State, Brazil, in 1985, it was deemed a spike disease with no foliar infections (Igarashi et al., 1986). However, after a few years of disease build up and epidemics in the southern and central regions of Brazil, foliar infections, especially on susceptible varieties, were observed. The wheat blast pathogen can attack all parts of the plant; however, the most serious infection occurs on the rachis because it bleaches the portion of the spike above the site of infection and no grains are formed. As a result, there are severe production losses, greater than 70% in early sown fields (Igarashi, 1990; Kohli et al., 2011). Even in areas considered endemic, wheat blast infections are extremely dependent on favourable environmental conditions conducive to the build-up of epidemics. The conditions are temperatures between 20°C and 30°C combined with high relative humidity during the heading and flowering stages of the crop. Such climatic conditions are very common during the El Niño years, making them most favourable for the development of an epidemic (Kohli et al., 2011; Fernandes et al., 2017).

As is the case with other spike diseases of wheat viz., *Fusarium* head blight, wheat blast disease needs to be managed through integrated control measures that must include the use of resistant wheat varieties and other cultural and chemical control practices to help reduce production losses. Yet, so far, all efforts to develop resistant varieties have stalled due to restricted genetic variability for resistance in the germplasm and lack of credible screening methodologies to use in the selection process.

Most commercial varieties released around the world derive their resistance to wheat blast from a single source, CIMMYT advanced line Milan, which was recently identified to carry the 2NS translocation derived from *Triticum ventricosum* (Kohli et al., 2011; Cruz et al., 2016). Despite the fact that a Japanese group (Zhan et al., 2008; Anh et al., 2015; Tagle et al., 2015) from the University of Kobe has identified four resistance genes (*Rmg2* and *Rmg3* from the bread wheat variety Thatcher, *Rmg7* from tetraploid *Triticum dicoccum* and *Rmg8* from bread wheat) against MoT strains, they are not effective against the pathogen strains found in South America (unpublished data). It must be pointed out that *Rmg2*, *Rmg3* and *Rmg7* conferred resistance to older MoT isolates tested between 1990 and 1992, but they are not effective against recent MoT isolates. The resistance of *Rmg8* remains to be tested with current MoT isolates and with natural field populations (Cruz & Valent, 2017).

8.2 ISOLATION, PURIFICATION AND CONSERVATION OF THE FUNGUS

While the methodologies pertaining to the isolation, purification and conservation of different fungi are well documented, some minor modifications to the standard procedures go a long way to promote efficiency in specific cases.

8.2.1 Isolation

In case of the wheat blast fungus, samples of wheat leaves and spikes showing characteristic symptoms of the disease can be isolated. In general, leaf symptoms are characterised by elliptical spots with a reddish-brown or dark border and a lighter centre (Reis et al., 2005). The characteristic symptoms on wheat spikes are the presence of greyish to black mycelia covering the point of infection on the rachis and the weed inflorescences show only necrotic lesions covered by a greyish mycelial growth.

The processing of disease samples in the laboratory for isolation follows a routine protocol proposed by French and Hebert (1980). A small portion of the diseased tissue is disinfected by immersion in 70% alcohol solutions for 30 seconds, followed by immersion in diluted sodium hypochlorite solution (55 g Cl/l) : water in 1 : 3 ratio) for one minute, then washed three times with sterile distilled water and dried on absorbent paper. The dried samples are seeded in petri dishes containing potato dextrose agar (PDA) culture medium (200 g potato, 20 g dextrose, 12 g/l agar) and incubated at 25°C for 5 days. After the incubation period, identification is carried out by observation under a microscope using the Barnett and Hunter (1998) code.

The second strategy is to isolate the fungus directly from the infected materials. In this case, the disinfected material is kept in a humid chamber for 24 hours and then with the aid of a magnifying glass, fungal conidia can be isolated and transferred to petri dishes containing an agar-water culture medium. After conidial germination (15 hours), they are transferred to an oat-agar medium for colony development (Cruz et al., 2009a).

8.2.2 Purification

After sporulation, 10 ml of sterile water is added to each petri dish containing the fungal culture and scraped in a laminar flow chamber. In order to obtain monosporic cultures, the above suspension is diluted three times to disperse the conidia, using 1 ml of the concentrated suspension with 9 ml of sterile water each time. A small amount (0.1 ml) of the diluted suspension is poured into a petri dish containing the PDA culture medium and dispersed using a sterile glass rod. It is imperative to observe the conidial germination under the stereomicroscope at regular intervals. At conidial germination, they are transferred individually to petri dishes containing the PDA culture medium (Gilchrist-Saavedra et al., 2005).

8.2.3 Conservation

For conservation, the conidial colonies are replicated on pieces of sterile filter paper, placed in petri dishes with the PDA culture medium and cultured for 8 days at 25°C. Subsequently, each piece is removed and placed in empty sterile petri dishes for drying and later storage at –18°C in Eppendorf tubes (Aricapa & Correa, 1994; Cruz et al., 2009).

8.3 FUNGAL SPORULATION AND PRODUCTION OF INOCULUM

8.3.1 Spore Production

In general, the fungus *M. oryzae* shows slow growth and has a lower capacity to sporulate under laboratory conditions, which hinders the production of large quantities of the inoculum (Soave et al., 1975). In addition, the media composition and the light regime affect sporulation significantly (Cruz et al., 2009). These authors tested eight culture media (rice wholemeal, oat, rye, barley, wholewheat flour, triticale, PDA and tomato juice) and three light regimes (dark, continuous light and 12 hours of light) to determine the frequency of sporulation. They maintained the petri dishes in a growth chamber for 10 days at 23°C–25°C. Their results showed that

TABLE 8.1
Average Sporulation of *Magnaporthe oryzae* in Different Culture Media

Culture media	Sporulation (conidia/ml)*
Oatmeal agar	4.1×10^6 a
Wheat spike agar	3.2×10^6 ab
V8 juice	2.1×10^6 ab
Wheat seed agar	2.7×10^5 b

*Averages with a common letter are not significantly different ($p > 0.05$).

regardless of the light regime, wholewheat flour favoured the highest sporulation (961,000 conidia/ml), while the lowest sporulation occurred in oatmeal, rice meal and PDA media.

Chavez and Kohli (2015) also tested four culture media: oatmeal agar, wheat seed agar, wheat spike agar and V8 juice, to determine the sporulation potential of local wheat pathotypes of *M. oryzae*. They used a protocol proposed by Marangoni et al. (2013), with a 12-hour light period and incubation at 28°C for 10 days, followed by crushing of the mycelium with a glass rod and reincubation under continuous light for 72 hours. Their results differed from those of Cruz et al. (2009) in that the agar-oat medium was found to be more favourable to wheat blast fungal sporulation (Table 8.1). This difference can be attributed to crushing the mycelium to promote sporulation.

8.3.2 Large-scale Inoculum Production

Any effort to produce a large-scale inoculum load starts with fungal isolates conserved on filter paper at –18°C in the *Magnaporthe* collection of Paraguay. Initially, a small portion of the pure culture is replicated in petri dishes containing an agar-oat culture medium. The inoculum for the greenhouse studies and field is also produced on the oatmeal-agar culture using the above-described methodology. In this case, the mycelium is crushed using an L-shaped glass rod, and the petri dishes containing fungal isolates are exposed to continuous fluorescent light for 72 hours to favour sporulation. The spores are removed using a paint brush and sterile distilled water (Marangoni et al., 2013). Although some researchers suggest adding Tween 80 (three drops per litre) to help dispersion and adhesion (Maciel et al., 2008; Cruz et al., 2012), this did not show any improvement in the efficiency of inoculation in our studies (unpublished data).

8.3.3 Impact of Inoculum Dosage

Artificial inoculations under the controlled conditions of the greenhouse and/or field are of critical importance in screening large numbers of breeding lines and identifying new sources of resistance to any disease. However, it also critical to use the

TABLE 8.2
Average Infection Score Caused by *Magnaporthe oryzae* on the Leaves of cv. Milan and cv. Canindé 11 at Different Concentrations of Inoculum

Inoculum concentration (conidios/ml)	cv. Milan	cv. Canindé 11
2.5×10^4	1 a*	3.3 b
7.5×10^4	1 a	3.3 b
1×10^5	1 a	3.6 b

*Infection score on the disease scale of 0–4: 0 = immune and 4 = susceptible. Averages with a common letter are not significantly different ($p > 0.05$).

correct inoculum dosage in order to generate a large variability of responses in the germplasm being tested and not miss small differences that can be important in certain combinations. In the case of wheat blast, several researchers have tried inoculum concentrations of between 2.5×10^4 conidia/ml and 1×10^6 conidia/ml (Maciel et al., 2008; Cruz et al., 2012; Marangoni et al., 2013).

In order to test the impact of inoculum concentration on wheat blast disease, Chávez et al. (2015) evaluated two wheat varieties, Milan (resistant) and Canindé 11 (susceptible), using three concentrations of inoculum (2.5×10^4, 7.5×10^4 and 1×10^5 conidia/ml). The results presented in Table 8.2 demonstrate that it is possible to differentiate between the resistant and the susceptible varieties at all three inoculum concentrations under study; however, it would be more useful to apply lower concentrations to identify germplasm with intermediate or moderate levels of resistance.

8.4 INOCULATION METHODOLOGY AND ENVIRONMENTAL CONDITIONS

8.4.1 Environmental Conditions

In an effort to develop an early warning system for the wheat blast disease, Cardoso et al. (2008) observed that the disease intensity varied significantly with the temperature and humidity (wetting period). Under controlled growth chamber experiments, they used a range of temperatures (10°C, 15°C, 20°C, 25°C, 30°C and 35°C) and durations of spike wetness (0, 5, 10, 15, 20, 25, 30, 35 and 40 hours) to record blast development on a susceptible cultivar BR23. Their results show that regardless of the temperature used, no blast symptoms developed when the wetting period was less than 10 hours. The highest blast intensity was observed at 30°C and increased when the duration of the wetting period increased. Even at 25°C, the blast development was lowest when the spike wetness was less than 10 hours, but with an increase in the wetting period to 40 hours at the same temperature, blast intensity exceeded 85% of infection.

These finding reconfirm the view that high humidity and/or a longer wetting period, prevalent in the El Niño years, are more critical to the development of wheat blast disease than the temperature regime.

8.4.2 Inoculation Methodologies

From the beginning, Brazilian researchers have used artificial inoculations on leaves to screen germplasm. However, following a large study, Chavez and Kohli (2018) showed conclusively that foliar infections, though easier to conduct, are not indicative of resistance in spikes. While spikes can be inoculated at any stage, in order to make the results uniform, inoculation is generally done at Stage 59 on Zadok's scale (spikes completely out of the flag leaf).

Artificial inoculation of the wheat blast under controlled greenhouse conditions can be achieved by spraying the spikes with a manual sprayer (three sprays per spike). Cruz et al. (2012) used a concentration of 2×10^4 conidia/ml at a rate of 1.2 ml of suspension per spike, and then covered each inoculated spike with a plastic bag for 48 hours. Chávez and Kohli (2018) used a concentration of 5×10^4 conidia/ml and kept the plants in a humid chamber for 24 hours or more (at 85 ± 5% humidity and 28 ± 2°C temperature) under normal daylight conditions. Later, the plants were moved to the greenhouse at the same temperature, more than 60% humidity and 12-hour light period for 14 days for the symptoms to become visible.

Field inoculations utilising sprinkler irrigation to maintain high levels of humidity are practiced in Brazil and Paraguay. However, due to the very high temperature regime during the day and higher evapotranspiration, the level of infection generally becomes irregular and is not a good indicator of germplasm resistance. In cases where a sprinkler is frequently used to maintain high humidity conditions, there are risks of blast infection getting mixed with *Fusarium* head blight and bacterial stripe infection. As a result, very early planting in autumn, i.e. March in the southern hemisphere, is practiced to serve as a source of inoculum to border the wheat germplasm planted at the normal date in the centre of the plot (Manoel Bassoi, personal communication, 2017). This protocol helps to increase the load of blast inoculum in the field but does not guarantee a uniform infection, unless it is accompanied by conducive environmental conditions.

In Paraguay, the germplasm for field inoculation is sown in 1-m long rows, 20 cm apart, early in the normal planting season (early April), aided by a sprinkler irrigation system to maintain high humidity during the critical heading and flowering stages of the crop. At the heading stage, approximately 20 spikes of similar age are marked per genetic material and spray inoculated individually with a suspension of 5×10^4 conidia/ml. Immediately after inoculation, each material is covered with a plastic-lined wooden box for 16 hours to maintain high humidity and promote infection. Later, the boxes are removed and sprinkler irrigation continues over the following days (Chávez et al., 2017). This protocol helps to achieve uniform infection and efficient screening of the targeted germplasm.

8.5 DISEASE SCORING

A disease evaluation scale, based on the percentage of spike infection, has been proposed by Maciel et al. (2013). In Paraguay, a modified version of this scale is used to assess large numbers of breeding lines and quantify the severity of wheat blast in the field (Figure 8.1). The modified 0–9 scale is based on the number of spikelets infected

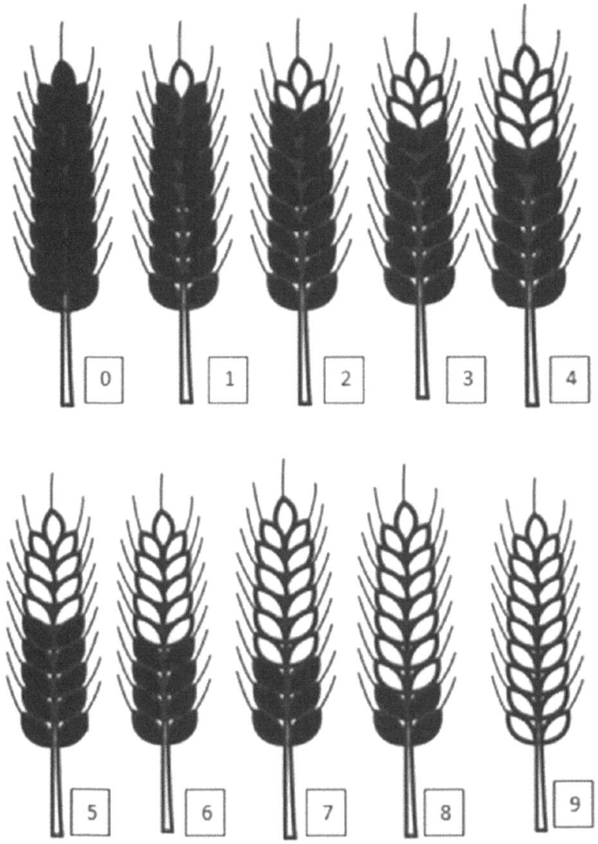

FIGURE 8.1 Modified Maciel scale used to score wheat blast in the field. 0 = no infection; 1 = 1 spikelet bleached (5% of spike bleached); 2 = 2–3 spikelet bleached (10% of spike bleached); 3 = 4–5 spikelet bleached (20% of spike bleached); 4 = 6–7 spikelet bleached (30% of spike bleached); 5 = 8–10 spikelet bleached (40% of spike bleached); 6 = 11–13 spikelet bleached (60% of spike bleached); 7 = 14–16 spikelet bleached (80% of spike bleached); 8 = 17–18 spikelet bleached (90% of spike bleached); 9 = complete bleached spike (100%).

in a spike, where zero (0) indicates no infection and nine (9) represents complete spike infection. It is important to emphasise that due to the variability in field infection, these evaluations need to be based on the population of at least 10 spikes per evaluated line. Considering that the inoculum pressure and/or climatic conditions at the time of infection can influence the point of infection, this scale serves to measure the disease severity in a germplasm but not necessarily its genetic resistance.

This critical aspect of wheat blast resistance scoring was included in an evaluation scale proposed by Tagle et al. (2015) (Figure 8.2). These researchers classify the blast lesions into five types:

0 = no infection; 1 = small lesions, <1.5 mm; 2 = intermediate lesions, <3 mm; 3 = mixture of green and white glumes, without apparent necrosis, caused by a hypersensitive reaction; and 4 = complete spike necrosis.

FIGURE 8.2 Wheat blast classification scale used to classify germplasm in the greenhouse (Tagle et al., 2015).

FIGURE 8.3 Screening for wheat blast infection under natural field conditions, Left = resistant line and right = susceptible line. IPTA Experimental Station, Yjovy, Paraguay.

This scale helps to classify the reaction types of a wheat genotype and it serves to identify the level of resistance or susceptibility in the germplasm under study. For field evaluations of the disease, a modified Maciel scale works well to estimate severity and production losses in general (Figure 8.3). However, the Tagle scale is more specific for research purposes to classify the genetic variability present in the germplasm and identify new sources of resistance to the wheat blast fungus.

Using the Tagle scale under artificially inoculated conditions in the greenhouse, and based on the average infection score of the spikes, Chávez and Kohli (2018) classify the genotypes into the following categories: immune (0), resistant (0.1–1), moderately resistant (1.1–2), moderately susceptible (2.1–3) and susceptible (3.1–4) (Figure 8.2).

8.6 SCREENING FOR WHEAT BLAST RESISTANCE

Most of the early work on screening for wheat blast resistance was done by foliar inoculations (Mehta & Baier, 1998; Mehta et al., 2001). While leaf reaction to the disease is useful to screen large numbers of genotypes in a short period of time, the resulting germplasm classification can be misleading in terms of wheat blast resistance in the spike. In order to emphasise the importance of spike infection, Chavez et al. (2017) conducted a comparative study to evaluate the blast reaction of 61 wheat genotypes in the vegetative and reproductive stages. Their results identified 35 of these 61 genotypes to be resistant in the vegetative stage. Of these 35 leaf-resistant genotypes, only 5 maintained their resistance in the reproductive stage, demonstrating that leaf resistance was not indicative of spike resistance (Table 8.3). It is interesting to observe that none of the leaf-susceptible genotypes gave resistant reaction in the spike.

Given a certain level of coincide between leaf and spike susceptibility, Chavez et al. (2017) suggested that the leaf reaction be used for preliminary screening to discard large numbers of susceptible germplasm and limit conducting spike infections to moderately resistant and resistant genotypes in the leaf.

Based on the knowledge cited above and in order to characterise the major wheat varieties grown in Paraguay, Chavez and Kohli (2018) evaluated the spike reaction of 32 varieties and 4 advanced lines to three isolates of the blast fungus. The results showed that only one cultivar, Canindé 1, carrying the 2NS segment from *Triticum ventricosa* via Milan, was resistant to the three isolates, while its sister line, CD 116, was resistant to one isolate but moderately resistant to the other two isolates. A Brazilian cultivar, TBIO Sintonía, also carrying the 2NS segment, was moderately resistant to the three blast isolates under study (Table 8.4). All other cultivars and

TABLE 8.3
Reaction Frequency of the 61 Wheat Genotypes to Leaf and Spike Infection of *M. Oryzae*

		Leaf reaction				
		Resistant	**Moderately resistant**	**Moderately susceptible**	**Susceptible**	**Total**
Spike reaction	Resistant	5	0	0	0	**5**
	Moderately resistant	2	0	1	1	**4**
	Moderately susceptible	5	4	3	1	**13**
	Susceptible	23	4	6	6	**39**
Total		**35**	**8**	**10**	**8**	**61**

TABLE 8.4
Classification of Major Wheat Varieties Grown in Paraguay against Three Isolates of *Magnaporthe* (synonym *Pyricularia*)

Material	Varietal classification based on average blast infection score in the greenhouse		
	P14-039	P14-025	P13-009
Caninde 1[1]	R	R	R
CANINDE 11[1]	S	S	S
CANINDE 13[1]	S	S	S
Caninde 21[1]	S	S	S
Caninde 3[1]	MS	S	MS
CANINIDE 12[1]	S	S	S
Itapúa 40[1]	S	S	S
Itapúa 70[1]	S	S	S
Itapúa 75[1]	S	MS	S
Itapúa 80[1]	S	S	S
Itapúa 85[1]	S	S	S
ITAPUA 75/WEBILL2[1]	S	S	S
ITAPUA 40/IAN 10[1]	S	S	S
E 92225/FUNDACEP30[1]	S	S	S
E 97034/ITAPUA 45[1]	S	S	S
BRS 208[2]	S	S	S
BRS 220[2]	S	MS	S
BRS Pardela[2]	S	S	S
BRS TANGARA[2]	S	S	S
CD 104[2]	S	S	S
CD 108[2]	S	S	MS
CD 116[2]	MR	MR	R
CD 150[2]	S	S	MS
CD 154[2]	S	S	S
FUNDACEP 6219[2]	MS	S	S
GRALHA AZUL[2]	S	S	S
IPR 144[2]	S	S	S
IPR CATUARA[2]	S	S	MS
IPR85[2]	S	S	S
LE 2331[2]	MS	S	S
QUARZO[2]	S	S	MS
TBIO IGUAZU[2]	S	MS	MR
TBIO MESTRE[2]	S	S	S
TBIO SINTONIA[2]	MR	MR	MR
TBIO TORUK[2]	MS	MS	MR

[1] Paraguayan cultivars and advanced lines.
[2] Brazilian cultivars seeded in Paraguay.

lines evaluated were moderately susceptible and susceptible. These results highlight the need to identify and introduce new sources of resistance in order to develop new blast-resistant cultivars in the future.

8.7 ANTICIPATORY SCREENING FOR WHEAT BLAST

For three decades, the wheat blast disease has been confined to the tropical South American regions of Bolivia, Brazil and Paraguay (Kohli et al., 2011). In Argentina, the disease has been limited to the border areas with Paraguay in the north; it has not spread to the major wheat-growing regions to the south. However, wheat blast identification in Bangladesh (Malaker et al., 2016) poses a serious concern to the adjoining countries with similar environmental conditions for the spread of disease. While the adaptation of the *Magnaporthe* fungus on wheat has been reported in this region, its infection on other grasses and weed species needs to be studied urgently in order to determine its variability in the near future.

As the South Asian region and other countries have an opportunity to control the impacts of wheat blast in its early phases, the screening protocols and methodologies presented in this chapter can be used to evaluate the wheat-breeding germplasm for resistance in safe containment facilities. It is essential that the pathogen is prevented from escaping into newer areas of production. While the plants can be grown under natural conditions for adequate development, they must be moved to a secure containment facility for inoculation and evaluation.

Any controlled screening in a new country needs to be conducted in a specially designed, high-level biosecurity structure that can guarantee safe and secure evaluation of exotic or high-impact pathogens. Moreover, entry to the biocontainment facility needs to be restricted to authorised personnel and a very strict protocol needs to be followed to work in such a facility. Depending on the availability of temperature and humidity control in the containment facility, the screening inside can be carried out in a glasshouse or plant growth chambers.

The other safe alternative for these countries is to collaborate with the South American programmes in testing their germplasm under field and controlled greenhouse conditions. Such an effort would render a quick evaluation of the germplasm base used in a breeding programme and allow anticipatory and focused breeding to be conducted in the country of origin. There are many examples of such international collaboration programmes conducted under the auspices of the International Maize and Wheat Research Centre (CIMMYT). The Wheat Blast Consortium was created to promote such collaboration and research among countries and must be utilised to keep production growing in new regions that are at risk from this menace.

REFERENCES

Anh, V. L., N. T. Anh, A. G. Tagle et al. "*Rmg8*, a New Gene for Resistance to *Triticum* Isolates of *Pyricularia oryzae* in Hexaploid Wheat." *Phytopathology* 105 (2015): 1568–72.

Aricapa, M. G. and V. F. Correa. *Integración de Fitopatología, mejoramiento y biología molecular para el desarrollo de resistencia estable al añublo del arroz (Pyricularia grisea).* Cali, Colombia: Centro Internacional de Agricultura Tropical, CIAT, 1994.

Barnett, H. L. and B. B. Hunter. *Illustrated Genera Imperfect Fungi*. 4th ed., 218. Saint Paul, MN: The American Phytopathological Society, 1998.

Cardoso, C. A., E. M. Reis and E. M. Moreira. "Development of a Warning System for Wheat Blast Caused by *Pyricularia grisea*." *Summa Phytopathologica* 34, no. 3 (2008): 216–21.

Chávez, A. and M. M. Kohli. "Hospederos alternativos de *Magnaporthe grisea* del trigo en Paraguay." *Investigación Agraria* 17, no. 1 (2015): 54–9.

Chávez, A. and M. M. Kohli. "Pathogenicity of *Magnaporthe oryzae* in Varieties and Wheat Lines Grown in Paraguay." *Revista Mexicana de fitopatología* 36, no. 2 (2018): 276–86. doi:10.18781/R.MEX.FIT.1712-3.

Chávez, A., C. Cazal and M. M. Kohli. "Diferencia en la reacción a *Pyricularia oryzae* de materiales de trigo en los estadios vegetativo y reproductivo." *Investigación Agraria* 19, no. 1 (2017): 56–63.

Chávez, A., C. Cazal, N. Bobadilla and M. M. Kohli. "Inoculación a campo de *Pyricularia oryzae* en genotipos de trigo." In *Pyricularia de trigo en Paraguay: Compendio de Investigación*, edited by A. Chávez and M. M. Kohli. Asunción, Paraguay: CAPECO, 2017.

Cruz, C. D and B. Valent. "Wheat Blast Disease: Danger on the Move." *Tropical Plant Pathology* 42, no. 3 (2017): 210–22.

Cruz, C. D., W. W. Bockus, J. P. Stack et al. "Preliminary Assessment of Resistance Among U.S. Wheat Cultivars to the *Triticum* Pathotype of *Magnaporthe oryzae*." *Plant Disease* 96 (2012): 1501–5.

Cruz, C. D., G. L. Peterson, W. W. Bockus et al. "The 2NS Translocation from *Aegilops ventricosa* Confers Resistance to the *Triticum* Pathotype of *Magnaporthe oryzae*." *Crop Science* 56 (2016): 990–1000.

Cruz, M. F., A. Prestes and J. L. Maciel. "Esporulação e crescimento vegetativo de *Pyricularia oryzae* em diferentes meios de cultura e regimes de luz." *Ciencia Rural* 39, no. 5 (2009a): 1562–4.

Cruz, M. F. A., J. L. N. Maciel, A. M. Prestes, E. A. S. Bombonatto, J. F. Pereira and L. Consoli. "Caraterizacao genética e fenotípica de isolados de *Pyricularia grisea* do trigo." *Tropical Plant Pathology* 34, no. 6 (2009b): 393–401.

Fernandes, J. M. C., M. Nicolau, W. Pavan et al. "A Weather-Based Model for Predicting Early Season Inoculum Build-Up and Spike Infection by the Wheat Blast Pathogen." *Tropical Plant Pathology* 42, no. 3 (2017): 230–7.

French, E. R. and T. T. Hebert. *Métodos de investigación fitopatológica*, 289. San José, CR: Instituto Interamericano de Ciencias Agrícolas, 1980.

Gilchrist-Saavedra, L., G. Fuentes-Davila, C. Martínez-Cano et al. *Guía práctica para la identificación de algunas enfermedades de trigo y cebada*. 2nd ed. Mexico D.F.: CIMMYT, 2005.

Igarashi, S. "Update on Wheat Blast (*Pyricularia oryzae*) in Brazil." In *Proceeding of the International Conference-Wheat for the Nontraditional Warm Areas*, edited by D. A. Saunders, 480–3. Mexico, D.F.: CIMMYT, 1990.

Igarashi, S., C. M. Utiamada, L. C. Igarashi, A. H. Kazuma and R. S. Lopes. "*Pyricularia* in Wheat. 1. Occurrence of *Pyricularia* sp. In Parana State." *Fitopatologia Brasileira* 11 (1986): 351–2.

Kohli, M. M., Y. R. Mehta, E. Guzman, L. de Viedma and L. Cubilla. "*Pyricularia* Blast: A Threat to Wheat Cultivation." *Czech Journal of Genetics and Plant Breeding* 47 (2011): S130–4. (Special Issue.)

Maciel, J. L., A. L. D. Danelli, C. Boretto and C. A. Forcellini. "Diagrammatic Scale for the Assessment of Blast on Wheat Spikes." *Summa Phytopathologica* 39, no. 3 (2013): 162–3.

Maciel, J. L., E. Paludo, M. Silva, P. L. Scheeren and E. Caierao. "Reação a brusone de genótipos de trigo do programa de melhoramento de Embrapa trigo no estádio de planta adulta." *Boletim de pesquisa e desenvolvimento* 64 (2008): 1–8. Paso Fondo, RS.

Malaker, P. K., N. C. D. Barma, T. P. Tiwari et al. "First Report of Wheat Blast Caused by *Magnaporthe oryzae* Pathotype *Triticum* in Bangladesh." *Plant Disease* 100, no. 11 (2016): 2330.

Marangoni, M., M. Nunes, N. Fonseca and Y. R. Mehta. "*Pyricularia* Blast on White Oats: A New Threat to Wheat Cultivation." *Tropical Plant Pathology* 38, no. 3 (2013): 198–202.

Mehta, Y. R. and A. Baier. "Variação patogênica entre isolados de *Magnaporthe grisea* atacando triticale e trigo no estado do Paraná." *Summa Phytopathologica* 24 (1998): 119–25.

Mehta, Y. R., C. A. A. Arias and J. F. F. Toledo. "Inheritance of Resistance to *Magnaporthe grisea* in Wheat." *Summa Phytopathologica* 2 (2001): 300–4.

Reis, E. M., R. T. Casa and C. A. Forcelini. "Doenças do trigo (*Triticum aestivum* L.)." In *Manual de Fitopatologia: doenças das plantas cultivadas.* 4 ed., vol. 2, edited by H. Kimati, L. Amorim, J. Rezende A. Bergamin and L. Camargo, 663. São Paulo: Ceres, 2005.

Soave, J., F. Galli and H. Kimati. "Estudo do crescimento vegetativo e da esporulação de *Pyricularia oryzae* Cavara, em diferentes meios de cultura." *Summa Phytopathologica, São Paulo* 1, no. 4 (1975): 258–66.

Tagle, A. G., I. Chuma and Y. Tosa. "*Rmg7*, a New Gene for Resistance to *Triticum* Isolates of *Pyricularia oryzae* Identified in Tetraploid Wheat." *Phytopathology* 105 no. 4 (2015): 495–9. doi:10.1094/PHYTO-06-14-0182-R.

Zhan, S. W., S. Mayama and S. Tosa. "Identification of Two Genes for Resistance to Triticum Isolates of *Magnaporthe oryzae* in Wheat." *Genome* 51 (2008): 216–21. doi:10.1139/G07/094.

9 Breeding for Wheat Blast Resistance

Xinyao He, Vikas Gupta, Naresh Kumar Bainsla, Aakash Chawade and Pawan Kumar Singh

CONTENTS

9.1 INTRODUCTION

Wheat is one of the most important cereal crops, sustaining food security across the globe and contributing significantly towards energy and protein requirements. Around the globe, wheat is cultivated in diverse climatic conditions ranging from subtropical to temperate. The demand for wheat and wheat products is increasing with the increase in population. It is projected that by 2050, the world's population will increase by 25%–33% from its current level with more than 65% of the increase occurring in Sub-Saharan Africa and South Asia. After the green revolution of the 1960s, wheat yield increased significantly, mainly attributed to improved semi-dwarf varieties and the judicious use of inputs. Recently, however, genetic gains have been about 1%, but the world's population is increasing at a greater pace. Further, wheat crop yields are also affected by various biotic and abiotic stresses causing serious yield losses. Additionally, diseases such as wheat blast (WB), which were confined to a limited area, are now emerging in other areas where it can be establish niche for virulent to highly virulent racial forms. The recent spread of WB in Bangladesh is an example. It poses a threat to the larger population residing in the tropical and subtropical regions of Asia, challenging the food security of about two-thirds of the entire human population.

Wheat blast, caused by *Magnaporthe oryzae* pathotype *Triticum* (MoT) B.C. Couch, is a disease of recent origin. It was first identified in Parana State, Brazil, in 1985 (Igarashi et al., 1986) and caused the loss of one-third of the region's wheat crop in 2009. WB was subsequently reported in Bolivia (Barea & Toledo, 1996), Paraguay (Viedma, 2005) and Argentina (Cabrera & Gutiérrez, 2007), and was recently reported in Bangladesh (Saharan et al., 2016; Malaker et al., 2016; Singh, 2017). This disease is the main reason for limiting the spread of wheat production in South America (Callaway, 2016).

The blast pathogen MoT causes bleaching of spikes, which is the main symptom of WB and can be easily recognised. Symptoms may appear on all the above ground parts of the plant, including the leaves, seed, rachis and peduncle. Grey-green, water-soaked lesions with necrotic borders appear on the leaves, whereas the pathogen produces dark brown discoloration on the rachis. The portion above the point of infection becomes bleached in the case of spike infection, which dries out quickly to give a bleached appearance (Igarashi, 1990; Islam et al., 2016; Malaker et al., 2016). If infection occurs during the grain-filling stage, then only small, shrivelled, lightweight and discoloured grains are produced. Highly susceptible cultivars can be infected at the seedling stage under optimum disease-developing conditions.

The pathogen can overwinter in seed, secondary hosts and crop residue. The yield losses caused by this pathogen vary from very low to 100% (Goulart & Paiva, 1990; Goulart et al., 1992). In Paraguay, the epidemic caused by this pathogen resulted in more than 70% losses in wheat yield (Viedma & Morel, 2002). Recently, with the appearance of the blast pathogen in Bangladesh, several WB-infected, government-owned farms have been burned (Callaway, 2016). The highest losses are incurred when the fungal infection occurs during heading at the base of the rachis, restricting the development of grains and killing the spike (Kohli et al., 2011). WB outbreaks are difficult to contain due to a number of factors including conducive weather conditions, variability in MoT virulence, lack of effective blast resistance in crops, poor chemical control as well as a lack of knowledge on different aspects of the plant–pathogen interaction (Cruz et al., 2011; Cruz & Valent, 2017). Fungicides have been used to control WB, but fungicide resistance has been reported in MoT populations, especially in Brazil, and this also adds to the burden of production costs (Ceresini et al., 2018). Therefore, the better and more eco-friendly option is to search and deploy host resistance to manage this disease. This chapter focuses on different breeding strategies to contain this devastating pathogen.

9.2 AVAILABLE RESISTANCE SOURCES

Currently, the biggest problem for blast resistance breeding is a lack of resistance sources that could be utilised in wheat breeding. Immediately after the first report of this disease in Brazil, several studies were conducted in order to identify a resistant germplasm (Barros et al., 1989; Igarashi, 1990; Goulart & Paiva, 1992). Several promising varieties were found, but all became susceptible in subsequent experiments (Urashima & Kato, 1994; Igarashi, 1990). Later, searches for resistance sources in Brazil led to the identification of several moderately resistant varieties such as BRS 49, BRS 120, BRS 220 and IAPAR 53 (Prestes et al., 2007). In Bolivia and Paraguay, however, no such screening activity was performed, and the identification of WB-resistant lines has solely relied on field observation over years. Several moderately resistant cultivars have been identified, e.g. Sausal CIAT, Motacu CIAT, Patuju CIAT and Urubo CIAT in Bolivia, and Caninde 1 and Itapua 75 in Paraguay (Buerstmayr et al., 2017). It has long been realised that Milan derivatives generally show a good level of resistance to WB, thus Milan-based resistant materials have been widely cultivated in Brazil, Bolivia and Paraguay (Kohli et al., 2011).

Later, it was found that the 2NS translocation from *Aegilops ventricosa* was the underlying resistance factor (Cruz et al., 2016), and most varieties with resistance to WB carry 2NS. However, it is dangerous to rely solely on 2NS resistance because once the 2NS-virulent isolates emerge and become prevalent, great blast epidemics and huge yield reductions are expected. In fact, 2NS-virulent MoT isolates have already evolved and are widely spread across Brazil, accounting for approximately 80% of the MoT isolates (Ceresini et al., 2018). So far, these isolates are known to have occurred in Paraguay and Bolivia too (Singh et al., 2016); fortunately, the MoT isolates prevalent in Bangladesh are non-virulent to 2NS (Singh et al., unpublished data). Nevertheless, considering the fast evolution of the 2NS-virulent isolates in South America, their appearance and spread in South Asia may not be avoided. Therefore, searching for non-2NS-based resistance has become the utmost priority. However, so far, no non-2NS line has been found to be resistant across various environments. Under low to moderate blast pressure or under certain environments, several non-2NS lines appear to be of low WB severity, such as Super 152, Quaiu #1 and Vorobey, but under high disease pressure, these lines could also suffer high WB infection. Currently, the search for new resistance sources is being conducted from a wider collection of wheat germplasm, and several promising non-2NS lines have been identified; however, further experiments are needed to confirm their resistance under wider environmental conditions.

Considering the likely limited resistance source in the primary gene pool, efforts have been made to search for new sources in the secondary and tertiary gene pool. Synthetic hexaploid wheat (SHW, AABBDD) is generated via crossing durum wheat (AABB) and *Aegilops tauschii* (DD) and subsequent chromosome doubling. Many SHW lines have shown consistent resistance in multiple nurseries, and since durum lines are generally susceptible to WB (Buerstmayr et al., 2017), the resistance is most likely derived from *Ae. tauschii*. Urashima and Kato (1994) screened a large number of *Aegilops* species for WB resistance and the results showed that *Ae. tauschii* and *Ae. umbellulata* were the only species resistant to all the isolates tested, providing theoretical support for the resistance of SHW to WB. It is also noteworthy that two Bolivian cultivars with moderate resistance, Motacu CIAT and Patuju CIAT, are SHW derivatives.

9.3 GENETIC BASIS OF RESISTANCE

The genetic basis of resistance has not yet been clearly elucidated because of the highly variable pathogen. However, both qualitative and quantitative genetic resistance have been identified in wheat against WB, but the former has been validated only at the seedling stage for the *Rmg* genes, which are discussed below. So far,

TABLE 9.1
List of Blast Resistance Genes Identified to Date

Resistance gene	Chromosome	Donor genotype	References
Rmg1 (Rwt4)	1D	Norin 4	Takabayashi et al. (2002)
Rmg2	7A	Thatcher	Zhan et al. (2008)
Rmg3	6B	Thatcher	Zhan et al. (2008)
Rmg4	4A	Norin 4	Nga et al. (2009)
Rmg5	6D	Red Egyptian	Zhan et al. (2008)
Rmg6 (Rwt 3)	1D	Norin 4	Vy et al. (2014)
RmgTd(t)	7B	*Triticum dicoccum* KU109 (Tat14)	Cumagun et al. (2014)
Rmg7	2A	*T. dicoccum* lines KU112(St17), 120(St24), KU122(St25)	Tagle et al. (2015)
Rmg 8 (2B)	2B	S-615	Anh et al. (2015)
2NS translocation	2AS/2NS	*Ae. ventricosa*, VPM1, Milan, etc.	Cruz et al. (2016)
RmgGR119	Unknown	GR119	Wang et al. (2018)

10 *Rmg* genes have been reported, including *Rmg1* through *Rmg8*, plus *RmgTd(t)* and *RmgGR119* (Table 9.1). These genes can be classified into two categories: non-host resistance genes and host resistance genes.

9.3.1 Non-host Resistance Genes, Conferring Resistance to Non-MoT Isolates of *Magnaporthe oryzae*

Apart from MoT, other pathotypes of *M. oryzae* infect various cereal and grass species, e.g. the *Oryza* pathotype (MoO) infecting rice, the *Avena* pathotype (MoA) infecting oat and the *Lolium* pathotype (MoL) infecting turf grasses. Based on the definition, these isolates do not infect wheat, but due to some rare mutation events, a few wheat lines might become susceptible to the non-MoT pathogens. A vivid example is the evolution of MoT from likely a MoL isolate in South America (Inoue et al., 2017).

***Rmg1* (*Rwt4*):** This gene corresponds to the avirulence gene *PWT4* in MoA, and was identified in wheat cultivar Norin 4 and located on chromosome 1D (Takabayashi et al., 2002; Hirata et al., 2005). *Rmg1* is a major factor in the resistance of Norin 4 to MoA and functions at all temperature ranges and all growth stages (Hau et al., 2007). However, *Rmg1* may not play an important role in wheat against *M. oryzae* pathogens, since it is mainly present in Japanese cultivars (Hirata et al., 2005). Furthermore, its corresponding avirulence gene *PWT4* has only been found in the oat and common millet isolates of *M. oryzae*, rather than the prevalent rice, wheat, foxtail millet and ryegrass isolates in South America (Hau et al., 2007).

***Rmg4* and *Rmg5*:** These two genes confer wheat resistance to *Digitaria* isolates of *M. grisea* (MoD), being located on wheat chromosomes 4A and 6D, respectively (Nga et al., 2009). The occurrence of these two genes in common wheat must be frequent, since few wheat lines are susceptible to MoD (Nga et al., 2009). It was also found that the two genes are temperature sensitive, i.e. they function normally at 22°C but their resistance breaks down at 26°C (Nga et al., 2009).

Rmg6 **(*Rwt3*):** This gene was mapped by Vy et al. (2014) on chromosome 1D in the wheat cultivars Chinese Spring, Shin-chunaga and Norin 4, conferring resistance to MoL. *Rmg6* recognises the avirulence gene *PWT3* in MoL and played an important role in the emergence of MoT isolates. This gene is prevalent in wheat cultivars, according to the survey conducted by Inoue et al. (2017), where 77% of the 499 wheat landraces of global origin had *Rmg6*. Unfortunately, the Brazilian cultivar Anahuac that was released in the early 1980s does not have this gene and thus can be infected by MoL-like isolates. Due to the wide cultivation of Anahuac, MoL-like isolates propagated in huge quantities on this cultivar and mutants with *pwt3*, which is virulent to wheat lines carrying *Rmg6*, soon evolved. The newly emerged isolates gained virulence to wheat cultivars and became MoT, which spread first within Brazil and then to neighbouring countries, causing great WB epidemics (Inoue et al., 2017).

***RmgTd(t)*:** This is regarded as a hidden resistance gene because it was not detected with the natural *M. oryzae* isolates, but with a white culture mutant from a cross between MoA and MoT. The mutant isolate was avirulent to all common wheat cultivars and the majority of tetraploid wheat, but it showed virulence to a few susceptible tetraploid wheat lines, thereby enabling the gene to be mapped on the 7B chromosome (Cumagun et al., 2014). Based on cytological analysis, *RmgTd(t)* is associated with the hypersensitive reaction of mesophyll cells (Cumagun et al., 2014).

9.3.2 Host Resistance Genes, Conferring Resistance to MoT Isolates

Rmg2 **and** ***Rmg3*:** These two genes were identified in the common wheat cultivar "Thatcher" and mapped on chromosomes 7A and 6B, respectively (Zhan et al., 2008). In the experiment, however, only seedling tests were performed and thus the effects of *Rmg2* and *Rmg3* on spike resistance remains unknown; however, according to Cruz and Valent (2017), MoT isolates collected after 2011 showed high aggressiveness and overcame the resistance of Thatcher; thus, it appears that these two genes have lost their resistance to newly emerged MoT isolates.

Rmg7 **and** ***Rmg8*:** *Rmg7* was identified in tetraploid *Triticum dicoccum* wheat accessions St24, St17 and St25, showing resistance to both leaf and head blast (Tagle et al., 2015). Later, the gene was mapped on chromosome 2A of tetraploid wheat (Anh et al., 2015). *Rmg8* was identified in the common wheat variety "S-615" and located on chromosome 2B, showing resistance in both leaf and spike tissues (Anh et al., 2015). Later, the avirulence gene corresponding to *Rmg8*, *AVR-Rmg8*, was cloned and found to encode a small protein containing a putative signal peptide. Interestingly, *AVR-Rmg8* was recognised not only by *Rmg8*, but also by *Rmg7*, suggesting that these two resistance genes are homoeologous and could be regarded as a single gene from the viewpoint of wheat breeding (Anh et al., 2018). Although *Rmg7* showed resistance to Br48, a MoT isolate collected in Brazil in the early 1990s, it lost its resistance to the recent MoT strains (Cruz & Valent, 2017). Therefore, it could also be inferred that *Rmg8* does not confer resistance to new MoT strains either because the two genes are recognised by the same avirulence gene (Anh et al., 2018).

2NS/2AS Translocation: This translocation was initially introduced into the common wheat variety "VPM1" in the 1960s, harbouring a 25–38 cM distal segment of the 2NS chromosome from *Ae. ventricosa* (Helguera et al., 2003). It has been extensively utilised due to the resistance genes resizing on the segment, e.g. *Lr37, Sr38* and *Yr17* for leaf, stem and stripe rust, respectively, *Rkn3* for root-knot nematodes and *Cre5* for the cereal cyst nematode (Cruz et al., 2016). Recently, Cruz et al. (2016) has associated this translocation with WB resistance in which wheat varieties with 2NS had 50.4%–80.5% less head blast than those without 2NS when inoculated with various MoT isolates. The famous WB resistance source Milan has the 2NS segment, as well as other well-known resistant or moderately resistant cultivars such as Urubo CIAT, Caninde #1 and Borlaug #100. It should be noted, however, that not all 2NS carriers show good resistance; it appears that the effectiveness of 2NS depends on genetic background (Cruz et al., 2016). Unlike *Rmg2, Rmg3, Rmg7* and *Rmg8* that have been overcome by recent MoT isolates, 2NS is still effective against most MoT isolates, at least in Bolivia and Bangladesh. Nevertheless, considering the large acreage of 2NS cultivars grown in South America and South Asia and the high variability of MoT isolates, it would be reasonable to predict the emergence of 2NS-virulent MoT strains. Actually, new MoT isolates have already shown higher aggressiveness against 2NS carriers compared to old MoT isolates (Cruz et al., 2016) and, according to unpublished data in Paraguay, 2NS-virulent MoT isolates have already been found (Singh et al., 2016).

***RmgGR119*:** This is the latest *Rmg* gene identified by Wang et al. (2018) in Albanian wheat accession GR119. The F_2 population derived from a cross between GR119 (resistant) and Chinese Spring (susceptible) were inoculated with the MoT isolate Br48, and the lines segregated in a 15 : 1 ratio for resistant vs. susceptible, suggesting that GR119 carries two resistance genes, of which one was found to be *Rmg8* and the other was tentatively named as *RmgGR119*. More work needs to be done on this gene regarding its chromosome location and effectiveness against new MoT isolates.

9.4 OPPORTUNITIES PROVIDED BY MODERN BIOTECHNOLOGY

The development of WB-resistant cultivars through conventional breeding takes 5–10 years, which is a major constraint. Therefore, modern genome editing techniques such as CRISPR/Cas9 would be useful in the development of blast-resistant wheat (Haque et al., 2018). Such a development could be achieved either by combining multiple R genes or deleting/disrupting susceptibility genes in the genome of a cultivar (Nekrasov et al., 2017; Peng et al., 2017). The latter approach would result in non-transgenic mutated wheat that could readily be released for practical application without having to follow strict biosafety guidelines (Langner et al., 2018; Haque et al., 2018). The WB susceptible genes have not been well characterised in wheat; therefore, the characterised orthologous susceptibility genes of rice could be targeted for genome editing (Haque et al., 2018). However, the pleotropic effects of the edited genes may affect the yield of the edited wheat. A targeted mutagenesis in wheat microspores using CRISPR/as9 has recently been established, which can be used to avoid these problems (Bhowmik et al., 2018). Wulff and Dhugga (2018) proposed a strategy of introducing blast resistance genes from wild progenitors of wheat, which could also be helpful in the rapid development of elite cultivars to confer multilayered disease resistance. The potential impact of gene editing on WB resistance has attracted much attention including in several recent research projects, where novel technologies have been included as a major component; for example, the project "Developing novel tools for breeding blast resistant wheat" at the Swedish University of Agricultural Sciences, in collaboration with CIMMYT and the Bangladesh Wheat and Maize Improvement Institute, and the project "Development of novel blast resistant wheat varieties for Bangladesh by genome editing" at the University of Exeter in collaboration with Bangabandhu Sheikh Mujibur Rahman Agricultural University. Once successfully demonstrated, these new technologies will significantly reduce the time for the release of resistant wheat varieties.

9.5 CONSORTIUM FOR WHEAT BLAST

In the last decades, wheat blast has been threatening South American countries, especially Brazil, Paraguay and Bolivia. However, there have been few active breeding programmes aimed at this disease, mainly due to the lack of a satisfactory resistance source and the limited understanding of host resistance genetics (Buerstmayr et al., 2017). WB was found in Bangladesh for the first time in 2016, and currently the main task is to screen South Asian wheat lines for WB resistance, so that promising lines can be utilised in future breeding programmes as WB resistance donors. The devastating WB epidemic in 2009 in South America drew global attention to this disease, resulting in the first international workshop on WB held in Paso Fundo, Brazil, in 2010. In the same year, a WB consortium was established that included researchers and breeders from CIMMYT, EMBRAPA, CAPECO and USDA/ARS, and advanced research institutes in the United States and Europe (Duveiller et al., 2016). In 2011, CIMMYT sent about 1,500 breeding lines to Brazil, Paraguay and Bolivia for blast evaluation, to exploit resistance sources to WB and to facilitate local breeders in utilising CIMMYT germplasm in their breeding programmes. No line turned out to be immune to this disease, but elite lines with good WB resistance were identified. The WB outbreak in Bangladesh in 2016 raised further global concern on this devastating disease and CIMMYT made an immediate response via (1) coordinating meetings/workshops and organising training courses on mitigating WB; (2) setting up precision phenotyping platforms (PPP) in Quirusillas and Okinawa, Bolivia, and Jashore, Bangladesh; and (3) enhancing WB screening for germplasm from CIMMYT and South Asia, as well as synthetic wheat lines and mapping populations. Subsequently, more research activities have been undertaken by CIMMYT and its collaborators on (1) genetics and breeding for WB resistance in wheat, (2) the fungal biology of the WB pathogen and (3) integrated WB management and seed systems. Several donors and NARS programmes have actively supported breeding for WB–resistant varieties, including the CGIAR Research Program on WHEAT, the Australian Centre for International Agricultural Research (ACIAR), Vetenskapsrådet – the Swedish Research Council, the Indian Council for Agricultural Research and the Bangladesh Wheat and Maize Research Institute.

9.6 CIMMYT'S RECENT EFFORTS ON WB RESISTANCE SCREENING AND BREEDING

Before the 2016 WB outbreak in Bangladesh, CIMMYT's actions on WB screening were achieved via collaborating with breeders in Brazil, Paraguay and Bolivia, i.e. CIMMYT sent germplasm to breeders for their evaluation and selection in local breeding nurseries. Because breeders made their selection based on multiple traits, very often varieties with superior WB resistance were discarded due to their poor performance in yield or spot blotch susceptibility or other traits. The WB outbreak in Bangladesh has driven the urgency to identify resistant germplasm that could be grown in Bangladesh as well as in India and Nepal as a pre-emptive measure. Under this scenario, CIMMYT set up two WB screening platforms in Bolivia and one in Bangladesh. In collaboration with its Bolivian partner, the National Institute for Agricultural and Forestry Research Innovation (INIAF), WB screening platforms were established in Quirusillas (highlands, wheat crop cycle from December to March) and Okinawa (lowlands, wheat crop cycle from May to September). In Bangladesh, a screening platform was set up in the south-western city of Jashore, at the regional agriculture research station of the Bangladesh Agriculture Research Institute (BARI). The wheat crop cycle runs from December to April in Jashore (lowland area), similar to the sowing and harvesting dates for much of southern Bangladesh where WB has been observed. The screening of intermediary materials from winter × spring and synthetic × spring crosses is still a challenge due to the small window of maturity cycle at these locations. Therefore, promising complementary or hidden factors which could provide alternatives to or supplement the 2NS segment from pre-breeding early generations is lacking at this point in time.

With the operation of the three screening platforms, results from multiple environments could be obtained in a year, significantly increasing the screening capacity and accelerating the release of WB resistant cultivars. A good example is the release of "BARI Gom 33" in Bangladesh. The performance of Bangladeshi materials was generally very poor in the WB screening platforms; however, a breeding line BAW1260 with the pedigree of "Kachu/Solala" showed consistently high resistance. Furthermore, this line has 30% higher grain zinc content than popular wheat varieties, making it very suitable for cultivation in South Asia where zinc deficiency is a severe problem. Based on these advantages, BAW1260 was released in Bangladesh as "BARI Gom 33" in October 2017, only one and a half years after the WB outbreak in the country.

Based on unpublished results from the three screening platforms, WB resistance is still heavily dependent on the 2NS translocation, as most of the resistant lines have 2NS. On the contrary, lines without 2NS showed either consistently high levels of susceptibility or highly unstable resistance, e.g. in experiment A it showed high resistance but in experiment B it turned out to be highly susceptible. The reasons for this high instability could be ascribed to (1) a lack of major resistance genes, making the lines appear to be resistant under low disease pressure but highly susceptible under high disease pressure; and (2) the influence of environmental factors and inaccurate screening techniques. It has long been observed that WB is sensitive to even minor environmental influences, e.g. wheat plants in the same field showed different WB severity due to soil compaction caused by tractor operations, and clear differences in WB severity in neighbouring fields sown within only a 1–2 day interval (Duveiller et al., 2016). Considering the gradually diminishing 2NS resistance, it is imperative to identify non-2NS resistance sources and utilise them in breeding programmes. CIMMYT is planning to collaborate with more countries in WB screening, in order to exploit non-2NS resistance genes as well as to identify useful resistant materials for pre-emptive breeding in the target countries.

In the last few years, crosses for WB resistance have been made at CIMMYT. The basic principle is to incorporate 2NS into elite lines as the backbone and, at the same time, to accumulate as many as possible of the yet unknown resistance genes. The utilisation of 2NS resistance has been successful as reflected in the very high frequencies of 2NS in nurseries such as the International Bread Wheat Screening Nursery (IBWSN) and the Semi-Arid Wheat Screening Nursery (SAWSN), where 80%–90% of lines have the translocation (Singh et al., unpublished data). This benefited from the long history of utilising *Yr17*, *Lr37* and *Sr38* that are located on the 2NS segment in CIMMYT wheat breeding programmes. On the other hand, non-2NS lines with moderate WB resistance have also been utilised to gradually accumulate minor resistance genes to diversify the *R* gene reservoir.

9.7 FUTURE PROSPECTS

So far, no large-scale breeding programme aimed specifically at WB resistance has been implemented, due to three main factors: (1) WB is a sporadic disease causing various yield losses across years; (2) host resistance is insufficient – even the 2NS carriers could be heavily infected under high pressure epidemics; and (3) WB resistance is not the only breeding target in all the affected countries. A vivid example of the latter is the cultivation of Motacu in Bolivia, which is moderately resistant to moderately susceptible to WB, but it is sown in much larger areas than many other resistant lines due to its early maturity (Vales et al., 2017). In a review article, Vales et al. (2017) proposed a range of breeding strategies under different host resistance scenarios, applying the experiences gained in breeding for rice blast resistance. Wheat would still be able to survive in WB-affected regions even if there was no host resistance at all (via a range of management strategies, which are not subjects of this chapter); however, according to current knowledge of WB resistance, there is at least partial resistance, so breeding for WB resistance should still be effective. Additionally, a large amount of wheat materials as well as their wild relatives have not been evaluated for WB resistance and thus it is still possible to find good resistance sources.

REFERENCES

Anh, V. L., N. T. Anh, A. G. Tagle et al. "*Rmg8*, a New Gene for Resistance to *Triticum* Isolates of *Pyricularia oryzae* in Hexaploid Wheat." *Phytopathology* 105, no. 12 (2015): 1568–72. doi:10.1094/PHYTO-02-15-0034-R.

Anh, V. L., Y. Inoue, S. Asuke et al. "*Rmg8* and *Rmg7*, Wheat Genes for Resistance to the Wheat Blast Fungus, Recognize the Same Avirulence Gene *AVR-Rmg8*." *Molecular Plant Pathology* 19, no. 5 (2018): 1252–6. doi:10.1111/mpp.12609.

Barea, G. and J. Toledo. "Identificación y zonificación de Pyricularia o brusone (*Pyricularia oryzae*) en el cutivo de trigo en el departamento de Santa Cruz." In *Centro de Investigación Agrícola Tropical. Informe Tecnico. Proyecto de Investigacion Trigo*, Santa Cruz de la Sierra, Bolivia, 1996. 76–86.

Barros, B. C., J. C. Felício, C. E. O. Camargo and A. W. P. Ferreira Filho. "Reação de cultivares de trigo a brusone em condições de campo." *Summa Phytopathologica* 15 (1989): 21.

Bhowmik, P., E. Ellison, B. Polley et al. "Targeted Mutagenesis in Wheat Microspores Using CRISPR/Cas9." *Scientific Reports* 8, no. 1 (2018): 6502. doi:10.1038/s41598-018-24690-8.

Buerstmayr, H., V. Mohler and M. Kohli. "Advances in Control of Wheat Diseases: Fusarium Head Blight, Wheat Blast and Powdery Mildew." In *Achieving Sustainable Cultivation of Wheat Volume 1: Breeding, Quality Traits, Pests and Diseases*, edited by P. Langridge, 345–70. Cambridge: Burleigh Dodds, 2017.

Cabrera, M. and S. Gutiérrez. "Primer registro de *Pyricularia grisea* en cultivos de trigo del NE de Argentina." *Depto. Protección Vegetal, Facultad de Ciencias Agrarias, UNNE*, 2007. www.agr.unne.edu.ar/images/documentos/extension/Res2007/SanVegetal/SanVegetal_06.pdf

Callaway, E. "Devastating Wheat Fungus Appears in Asia for First Time." *Nature* 532, no. 7600 (2016): 421–2.

Ceresini, P. C., V. L. Castroagudín, F. Á. Rodrigues et al. "Wheat Blast: Past, Present, and Future." *Annual Review of Phytopathology* 56 (2018): 427–56. doi:10.1146/annurev-phyto-080417-050036.

Cruz, C. D. and B. Valent. "Wheat Blast Disease: Danger on the Move." *Tropical Plant Pathology* 42, no. 3 (2017): 210–22. doi:10.1007/s40858-017-0159-z.

Cruz, C. D., G. L. Peterson, W. W. Bockus et al. "The 2NS Translocation from *Aegilops ventricosa* Confers Resistance to the *Triticum* Pathotype of *Magnaporthe oryzae*." *Crop Science* 56 (2016): 990–1000. doi:10.2135/cropsci2015.07.0410.

Cruz, M. F. A da, A. P. C. Diniz, F. A. Rodrigues and E. G. de Barros. "Foliar Application of Products on the Reduction of Blast Severity on Wheat." *Tropical Plant Pathology* 36, no. 6 (2011): 424–8.

Cumagun, C. J. R., V. L. Anh, T. T. P. Vy et al. "Identification of a Hidden Resistance Gene in Tetraploid Wheat Using Laboratory Strains of *Pyricularia oryzae* Produced by Backcrossing." *Phytopathology* 104, no. 6 (2014): 634–40. doi:10.1094/PHYTO-04-13-0106-R.

Duveiller, E., X. He and P. K. Singh. "Wheat Blast: An Emerging Disease in South America Potentially Threatening Wheat Production." In *The World Wheat Book Vol 3*, edited by A. Bonjean and M. van Ginkel, 1107–22. Paris: Lavoisier, 2016.

Goulart, A. C. P. and F. de A. Paiva. "Transmission of *Pyricularia oryzae* by Wheat (*Triticum aestivum*) Seeds." *Fitopatologia Brasileira* 15, no. 4 (1990): 359–62.

Goulart, A. C. P. and F. de A. Paiva. "Incidência da brusone (*Pyricularia oryzae*) em diferentes cultivares de trigo (*Triticum aestivum*) em condições de campo." *Fitopatologia Brasileira* 17, no. 3 (1992): 321–5.

Goulart, A. C. P., F. de A. Paiva and N. Mesquita. "Perdas em trigo (*Triticum aestivum*) causadas por *Pyricularia oryzae*." *Fitopatologia Brasileira* 17 (1992): 115–17.

Haque, E., H. Taniguchi, M. M. Hassan et al. "Application of CRISPR/Cas9 Genome Editing Technology for the Improvement of Crops Cultivated in Tropical Climates: Recent Progress, Prospects, and Challenges." *Frontiers in Plant Sciences* 9 (2018): 617. doi:10.3389/fpls.2018.00617.

Hau, V. T. B., K. Hirata, J. Murakami, H. Nakayashiki, S. Mayama and Yukio Tosa. "*Rwt4*, a Wheat Gene for Resistance to *Avena* Isolates of *Magnaporthe oryzae*, Functions as a Gene for Resistance to *Panicum* Isolates in Japan." *Journal of General Plant Pathology* 73, no. 1 (2007): 22–28.

Helguera, M., I. A. Khan, J. Kolmer, D. Lijavetzky, L. Zhong-qi and J. Dubcovsky. "PCR Assays for the Lr37-Yr17-Sr38 Cluster of Rust Resistance Genes and Their Use to Develop Isogenic Hard Red Spring Wheat Lines." *Crop Science* 43, no. 5 (2003): 1839–47.

Hirata, K., Y. Tosa, H. Nakayashiki and S. Mayama. "Significance of *PWT4–Rwt4* Interaction in the Species Specificity of *Avena* Isolates of *Magnaporthe oryzae* on Wheat." *Journal of General Plant Pathology* 71, no. 5 (2005): 340–4.

Igarashi, S. "Update on Wheat Blast (*Pyricularia oryzae*) in Brazil." In *Wheat for the Nontraditional Warm Areas. A Proceedings of the International Conference*, edited by David A. Saunders, 480–5. Foz do Iguaçu, Brazil: CIMMYT, 1990.

Inoue, Y., T. T. P. Vy, K. Yoshida et al. "Evolution of the Wheat Blast Fungus Through Functional Losses in a Host Specificity Determinant." *Science* 357, no. 6346 (2017): 80–3. doi:10.1126/science.aam9654.

Islam, T., D. Croll, P. Gladieux et al. "Emergence of Wheat Blast in Bangladesh was Caused by a South American Lineage of *Magnaporthe oryzae*." *BMC Biology* 14, no. 1 (2016): 84. doi:10.1101/059832.

Kohli, M. M., Y. R. Mehta, E. Guzman, L. De Viedma and L. E. Cubilla. "Pyricularia Blast: A Threat to Wheat Cultivation." *Czech Journal of Genetics and Plant Breeding* 47 (2011): S130–4.

Langner, T., S. Kamoun and K. Belhaj. "CRISPR Crops: Plant Genome Editing Toward Disease Resistance." *Annual Review of Phytopathology* 56 (2018): 479–512. doi:10.1146/annurev-phyto-080417-050158.

Lgarashi, S., C. M. Utiamada, L.C. Lgarashí, A. H. Kazuma and R. S. Lopes. "*Pyricularia* em trigo. 1. Ocorrencia de *Pyricularla* sp. no estado do Parana." *Fitopatologia Brasileira* 11 (1986): 351–2.

Malaker, P. K., N. C. D. Barma, T. P Tewari et al. "First Report of Wheat Blast Caused by *Magnaporthe oryzae* Pathotype *Triticum* in Bangladesh." *Plant Disease* 100 (2016): 2330.

Nekrasov, V., C. Wang, J. Win, C. Lanz, D. Weigel and S. Kamoun. "Rapid Generation of a Transgene-Free Powdery Mildew Resistant Tomato by Genome Deletion." *Scientific Reports* 7, no. 1 (2017): 482. doi:10.1038/s41598-017-00578-x.

Nga, N. T. T., V. T. B. Hau and Y. Tosa. "Identification of Genes for Resistance to a *Digitaria* Isolate of *Magnaporthe grisea* in Common Wheat Cultivars." *Genome* 52 (2009): 801–9. doi:10.1139/G09-054.

Peng, A., S. Chen, T. Lei et al. "Engineering Canker-Resistant Plants Through CRISPR/Cas9-Targeted Editing of the Susceptibility Gene *CsLOB1* Promoter in Citrus." *Plant Biotechnology Journal* 15, no. 12 (2017): 1509–19. doi:10.1111/pbi.12733.

Prestes, A. M., P. F. Arendt, J. M. C. Fernandes and P. L. Scheeren. "Resistance to *Magnaporthe grisea* Among Brazilian Wheat Genotypes." In *Wheat Production in Stressed Environments*, edited by H. T. Buck, J. E. Nisi and N. Salomon, 119–23. Dordrecht: Springer, 2007.

Saharan, M. S., S. C. Bhardwaj, R. Chatrath, P. Sharma, A. K. Choudhary and R. K. Gupta. "Wheat Blast Disease – An Overview." *Journal of Wheat Research* 8, no. 1 (2016): 1–5.

Singh, D. P. "Wheat Blast: A New Challenge to Wheat Production in South Asia." *Indian Phytopathology* 70, no. 2 (2017): 169–77. doi:10.24838/ip.2017.v70.i2.70609.

Singh, R. P., P. K. Singh, J. Rutkoski et al. "Disease Impact on Wheat Yield Potential and Prospects of Genetic Control." *Annual Review of Phytopathology* 54, no. 1 (2016): 303–22. doi:10.1146/annurev-phyto-080615-095835.

Tagle, A. G., I. Chuma and Y. Tosa. "*Rmg7*, A New Gene for Resistance to *Triticum* Isolates of *Pyricularia oryzae* Identified in Tetraploid Wheat." *Phytopathology* 105, no. 4 (2015): 495–9. doi:10.1094/PHYTO-06-14-0182-R.

Takabayashi, N., Y. Tosa, H. S. Oh and S. Mayama. "A Gene-for-Gene Relationship Underlying the Species-Specific Parasitism of *Avena/Triticum* Isolates of *Magnaporthe grisea* on Wheat Cultivars." *Phytopathology* 92, no. 11 (2002): 1182–8.

Urashima, A. S. and H. Kato. "Varietal Resistance and Chemical Control of Wheat Blast Fungus." *Summa Phytopathologica* 20 (1994): 107–12.

Vales, M., T. Anzoátegui, B. Huallpa and M. I. Cazon. "Review on Resistance to Wheat Blast Disease (*Magnaporthe oryzae Triticum*) from the Breeder Point-of-View: Use of the Experience on Resistance to Rice Blast Disease." *Euphytica* 214, no. 1 (2017): 1. doi:10.1007/s10681-017-2087-x.

Viedma, L. "Wheat Blast Occurrence in Paraguay." *Phytopathology* 95 (2005): 152.

Viedma, L. Q. and W. Morel. *Añublo o Piricularia del Trigo (Wheat Blast): MAG/DIA/CRIA*. Programa de Investigación de Trigo, CRIA, Capitán Miranda, Itapúa, 2002.

Vy, T. T. P., G. S. Hyon, N. T. T. Nga, Y. Inoue, I. Chuma and Y. Tosa. "Genetic Analysis of Host-Pathogen Incompatibility Between *Lolium* Isolates of *Pyricularia oryzae* and Wheat." *Journal of General Plant Pathology* 80, no. 1 (2014): 59–65. doi:10.1007/s10327-013-0478-y.

Wang, S., S. Asuke, T. T. P. Vy et al. "A New Resistance Gene in Combination with *Rmg8* Confers Strong Resistance Against *Triticum* Isolates of *Pyricularia oryzae* in a Common Wheat Landrace." *Phytopathology* 108, no. 11 (2018): 1299–306. doi:10.1094/PHYTO-12-17-0400-R.

Wulff, B. B. H. and K. S. Dhugga. "Wheat: The Cereal Abandoned by GM." *Science* 361, no. 6401 (2018): 451–2. doi:10.1126/science.aat5119.

Zhan, S. W., S. Mayama and Y. Tosa. "Identification of Two Genes for Resistance to *Triticum* Isolates of *Magnaporthe oryzae* in Wheat." *Genome* 51, no. 3 (2008): 216–21. doi:10.1139/G07-094.

10 Integrated Management of Wheat Blast Disease

Man Mohan Kohli, Cinthia Cazal and Alice Chavez

CONTENTS

10.1 INTRODUCTION

Wheat blast (WB) disease is caused by *Magnaporthe oryzae* (synonym *Pyricularia oryzae*) (Couch & Kohn, 2002). Several authors have referred to it being caused by *Magnaporthe grisea* (T.T. Hebert) M.E. Barr (anamorfo *Pyricularia grisea* Sacc.). Recently, Cruz and Valent (2017) have attributed its specificity on wheat to a subpopulation of *Magnaporthe oryzae*known as *Triticum* pathotype (MoT). In 2016, based on phylogenetic analyses, Castroagudín et al. (2016) identified a group of isolates obtained from wheat and other Poaceae hosts which they named as a new species *Pyricularia* *graminis tritici*. However, Gladieux et al. (2018) point out that despite a certain degree of specificity to different hosts in the *M. oryzae* isolates, which is a product of regular gene flow among them, it is not enough to group them as separate species.

Among the new diseases attacking wheat, WB is one of the most serious, causing spike infection and severe production losses in the tropical region of South America (Kohli et al., 2011). Its recent identification in Bangladesh in 2016 (Malaker et al., 2016; Callaway, 2016) has caused international alarm that it could affect the food security of millions in South Asia and other regions.

Since the first identification of WB in Londrina, state of Parana, Brazil, in 1985 (Igarashi et al., 1986), it has spread rapidly to other parts of Brazil, causing serious crop losses in the central and central south wheat-growing regions (Picinini & Fernandes, 1989; Goulart et al., 1990; Anjos et al., 1996). After many years of absence, its infection reappeared in the southern state of Rio Grande do Sul during the wet year of 2014 (Embrapa Trigo, 2014). The disease is also prevalent in other tropical wheat-growing regions of the Southern Cone, such as northern Argentina, the Santa Cruz region and the lowlands of Bolivia and Paraguay (Kohli et al., 2011).

10.2 HOST RANGE AND DISEASE SEVERITY

Over the years, *M. oryzae* has been shown to attack a wide range of cereal hosts viz. barley, oats, rye and triticale (Mehta & Baier, 1998; Urashima et al., 2004b; Marangoni et al., 2013) and a large number of grassy weeds present in and around wheat fields (Chávez & Kohli, 2015).

While the grass weeds and ground stubble provide a continuous source of inoculum year after year, WB infection is extremely dependent on the weather conditions, especially high humidity (or free water in the form of dew) and high temperatures. Severe epidemics of WB are more common in the Niño years that characterise such weather patterns. Considering that both WB and *Fusarium* head blight (FHB) disease favour similar environmental conditions, it has become common to observe mixed infections of the two diseases in the field in Paraguay.

Toledo and Escobar (2002) recorded that fungal sporulation of WB can occur between 10°C and 35°C and conidial germination is favoured when temperatures are between 22°C and 28°C; the optimum temperature for sporulation of the disease is around 28°C. These authors also observed conidial production to start at 93% relative humidity and favoured long wet periods. This was further confirmed by Cardoso et al. (2008), who determined that the highest WB severity was observed at 30°C and increased with the duration of the crop wetting period. These authors also observed that under controlled conditions, regardless of the temperature regime, no symptoms occurred when the wetting period was less than 10 hours. However, with a 40-hour continuous wetting period, severe blast intensity (over 85%) was observed even at 25°C. These are fairly common climatic conditions during the winter season in a Niño year, which favour WB infections.

Since the first observation of the disease in 1985, the wheat blast disease has been observed to be most destructive in its infection on the rachis and the spike. However, with the passage of time, WB has adapted to other plant parts and is now capable of attacking the foliar parts of the crop. The symptoms on the leaves and sheath can vary in size, form and colour depending on the stage of the crop and the genetic resistance of the variety. In general, WB infections cause brown or dark brown elliptical lesions with pointed ends. The central part of the lesion varies in colour from light brown to white bordered by red-brown or dark brown areas. In highly susceptible varieties, severe leaf infections can occur early and result in coalescing of the lesions, thereby giving an appearance of burnt leaves and sheath.

The spike infection is characterised by partial or total whitening (drying) of the spike above the point of infection or lesion on the rachis, characterised by a bright dark brown to black necrotic lesion. Although the disease can infect at any time after the emergence of the spike, the severest damage is caused early in the heading or flowering stage of the crop. Early spike infection aborts grain filling above the point of infection and this portion of the spike dries out giving the appearance of a white head. An infection at the base of the rachis affects the whole spike and white heads are visible all over the field. Post grain-filling infections result in varying degrees of grain shrivelling and lower production losses.

Crop losses caused by WB disease vary from field to field and year to year depending on the climatic conditions and the agronomic practices utilised by the farmers. In commercial crops, production losses from less than 10% to 100% have been observed in an epidemic year (Kohli et al., 2011). Severe production losses, especially in the epidemic years, were the reason for a reduction in the wheat area in the State of Mato Grosso do Sul, Brazil, from 430,000 ha in 1987 to less than 30,000 ha in 1997, and in Santa Cruz, Bolivia, from over 110,000 ha in 1997 to almost 25,000 ha in 2003, mentioned on the websites CONAB (2015) and CAO (2014). During the 2014 and 2015 crop seasons, wheat farmers in Paraguay and the adjoining state of Parana, Brazil, suffered severe crop losses of up to 50% from wheat blast disease, making the crop economically unattractive. Over the past three years, WB infections have been irregular and of minor importance in the Southern Cone region; however, its recent identification in South Asia has the potential to spread globally under the scenario of global warming or climate change.

Given the limited sources of resistance available to develop resistant varieties, integrated disease control measures are essential for the efficient control of WB disease. The objective of an integrated approach is to reduce the abundance of inoculum, impede or inhibit infection and/or decrease disease development under favourable conditions.

10.3 CULTURAL MANAGEMENT

10.3.1 Elimination of Secondary Hosts and Grassy Weeds

Considering that entire wheat crop in the Southern Cone region is seeded in stubble, using conservation tillage practices, the crop rotation scheme takes unprecedented measures to avoid crop stubble that favours WB infection. However, black oats, which are commonly used as ground cover for the summer soybean crop, and a pathogen host, complicate this situation. Similarly, volunteer wheat plants from the previous crop and grassy weeds in and around the field remain very important suspects as the primary source of *M. oryzae* inoculum (Table 10.1). Their elimination

TABLE 10.1
Secondary and Grassy Hosts of *Magnaporthe* spp.

Host	Species reported[a]	Genotype group	References
Anthoxanthum odoratum	*M. oryzae*	ND	Klaubauf et al. (2014)
Avena sativa	*M. oryzae*	Lolium[b]	Inoue et al. (2017)
Avena strigosa	ND	Lolium[b]	Chávez and Kohli (2018b)
Brachiaria mutica	*M. oryzae*	Brachiaria[b]	Klaubauf et al. (2014); Gladieux et al. (2018)
Brachiaria distachya	*M. oryzae*	Brachiaria[b]	Gladieux et al. (2018)
Brachiaria plantaginea	ND	ND	Prabhu et al. (1992); Urashima et al. (1993)
Brachiaria sp.	ND	ND	Chávez and Kohli (2015)
Bromus catharticus	ND	Lolium/Triticum[b]	Chávez and Kohli (2015)
Bromus tectorum	*M. oryzae*	Lolium[c]/Triticum[c]	Pieck et al. (2017)
Cenchrus echinatus	*M. pennisetigena*	ND	Klaubauf et al. (2014); Wang et al. (2019); Chávez and Kohli (2018b)
Cloris gayana	ND	ND	Chávez and Kohli (2015)
Cynodon dactylon	*M. oryzae*	Cynodon[c]	Klaubauf et al. (2014); Castroagudín et al. (2016); Yasuhara-Bell et al. (2018)
Cyperus diffusus	ND	ND	Chávez and Kohli (2018b)
Cyperus rotundifolia	*Magnaporthe* sp.	ND	Yasuhara-Bell et al. (2018)
Digitaria ciliaris	*M. grisea*	ND	Klaubauf et al. (2014)
Digitaria horizontales	*M. grisea*	ND	Klaubauf et al. (2014); Chávez and Kohli (2015)
Digitaria insularis	ND	ND	Chávez and Kohli (2018b)
Digitaria sanguinalis	*M. grisea*	ND	Klaubauf et al. (2014); Gladieux et al. (2018)
Digitaria smutsii	*M. grisea*	ND	Klaubauf et al. (2014)
Echinochloa colona	*M. oryzae*	Echinochoa[c]	Klaubauf et al. (2014); Yasuhara-Bell et al. (2018)
Echinochloa crusgalli	*M. grisea/M. oryzae*	ND	Klaubauf et al. (2014)
Eleusine coronata	*M. oryzae*	Eleusine[b]	Gladieux et al. (2018)
Eleusine indica	*M. oryzae*	Eleusine[b]	Klaubauf et al. (2014); Gladieux et al. (2018); Chávez and Kohli (2018b)
Elionorus candidus	*M. oryzae/P. gramini-tritici*		Castroagudín et al. (2016)
Eragrotis curvula	*M. oryzae*	Eragrotis[b]	Klaubauf et al. (2014); Islam et al. (2016)
Eragrotis spp.	*M. oryzae*	Digitaria[c]	Yasuhara-Bell et al. (2018)
Eriochloa villasa	*M. oryzae*	ND	Klaubauf et al. (2014)
Festuca elalior	*M. oryzae*	ND	Klaubauf et al. (2014)
Fetuca arundinaceum	*M. oryzae*	Lolium[b]	Gladieux et al. (2018)

(Continued)

TABLE 10.1 (CONTINUED)
Secondary and Grassy Hosts of *Magnaporthe* spp.

Host	Species reported[a]	Genotype group	References
Hordeum vulgare	*M. oryzae*	Oryza[b]	Islam et al. (2016); Gladieux et al. (2018)
Hyparrhenia ruffa	ND	ND	Prabhu et al. (1992); Urashima et al. (1993)
Leersia hexandra	*M. oryzae/ Magnaporthe* sp.	Leersia[c]	Klaubauf et al. (2014); Yasuhara-Bell et al. (2018)
Leptochlo chimensis	*M. oryzae*	Leptocloa[c]	Klaubauf et al. (2014); Yasuhara-Bell et al. (2018)
Lolium hybridum	*M. oryzae*	ND	Klaubauf et al. (2014)
Lolium multiflorum	*M. oryzae*	Lolium[b]	Chávez and Kohli (2015); Gladieux et al. (2018)
Lolium perenne	*M. oryzae/M. grisea*	Lolium[b]	Klaubauf et al. (2014); Gladieux et al. (2018)
Paspalum distichum	*M. oryzae*	ND	Klaubauf et al. (2014)
Paspalum sp.	*M. oryzae*	ND	Klaubauf et al. (2014)
Panicum repens	*M. oryzae*	Panicum[c]	Klaubauf et al. (2014); Yasuhara-Bell et al. (2018)
Pennisetum americanum	*M. penniseticola*	ND	Gladieux et al. (2018)
Pennisetum typhoides	*M. penniseticola*	ND	Klaubauf et al. (2014)
Pennisetum setosum	ND	ND	Prabhu et al. (1992); Urashima et al. (1993)
Phalaris arundinacea	*M. oryzae*	ND	Klaubauf et al. (2014)
Phleum pratense	*M. oryzae*	ND	Klaubauf et al. (2014)
Rhynchelytrum roseum	ND	ND	Urashima et al. (1993); Chávez and Kohli (2018b)
Rhynchelytrum repens	*M. oryzae/P. graminis-tritici*	ND	Castroagudín et al. (2016)
Rottboellia exalta	*M. oryzae*	ND	Klaubauf et al. (2014)
Urochloa brizantha	*M. oryzae/P. gramini-tritici*	ND	Castroagudín et al. (2016)
Setaria faberi	*M. oryzae*	ND	Yasuhara-Bell et al. (2018)
Setaria viridis	*M. oryzae*	Setaria[b]	Klaubauf et al. (2014); Gladieux et al. (2018)
Setaria italica	*M. oryzae*	ND	Yasuhara-Bell et al. (2018)
Stenotaphrum secundatum	*M. oryzae*	Stenotaphrum[c]	Klaubauf et al. (2014); Yasuhara-Bell et al. (2018)
Zea mays	*M. oryzae*	ND	Klaubauf et al. (2014)

[a] Polyphasic identification (morphological and molecular).
[b] Genotype group reported in Gladieux et al. (2018).
[c] Genotype group reported in Yasuhara-Bell et al. (2018).
ND: Not determined.

is considered essential in reducing the initial inoculum load, thereby avoiding early onset of infection (Toledo & Escobar, 2002). As many weeds and grasses in the pastures are perennial and also potential WB pathogen hosts, this strategy has limited value in the Southern Cone region, but may work better in other parts of the world where agriculture and animal husbandry do not overlap.

10.3.2 The Use of Clean Seed

Several authors have confirmed the seed transmission of wheat blast disease whereby the pathogen can move from seed to coleoptile and primary leaves on the seedlings (Tanaka et al., 1988; Goulart et al., 1989). However, the efficiency of transmission can vary from 20% to 47.3% of the infected seed (Goulart et al., 1990). In other words, if the seedling emergence is not affected by the pathogen, a seed lot with 21% natural WB infection can result in 400,000 infection foci/ha. The assertion of Islam et al. (2016) that wheat blast in Bangladesh was caused by a South American lineage of *M. oryzae* strongly suggests that infected seed was behind the introduction of this disease into a newer region.

10.3.3 Date of Seeding

The practice of early seeding of wheat has been identified as one of the most important factors in the resurgence of WB in the Southern Cone region. In most cases, seeding wheat early in autumn (March or early April in the Southern Hemisphere) allows the heading and flowering period to coincide with warm and humid climatic conditions, exposed to severe WB infections except in the dry years.

These crops suffer most damage from WB on a year-to-year basis. Delayed planting, after 10 April, was able to cut production losses in the epidemic areas of Parana, Brazil, because the heading coincided with a non-conducive drier environment (Mehta et al., 1992). Similarly, the state government of the Department of Santa Cruz, Bolivia, banned farmers from planting wheat crop before 20 April in order to reduce the damage caused by WB (PROTRIGO, 2002). Oliveira Coelho et al. (2016) established a direct and significant interaction between different sowing dates and agronomic characteristics such as grain yield, thousand kernel weight and the disease index for WB. They confirmed that late sowing results in lower incidence of WB disease in this region. Even when seeding wheat in the month of May is recommended in the lowlands of Bolivia and Paraguay to avoid WB, variable levels of infections can still be observed depending on the climatic conditions of the year (Kohli et al., 2010). Multiple dates of seeding and varieties generally help reduce crop losses.

10.3.4 Seeding Density

A seeding density higher than the recommended rate for a variety or date of planting can lead to strong competition in the vegetative stage and result in earlier than normal heading and flowering. This is especially true for high tillering varieties which are grown under optimum conditions. Furthermore, a closed canopy of a crop can also create a micro-climate favourable for pathogen multiplication, resulting in higher infections. Considering that poor management practices such as lack of crop rotation and climatic factors such as a high temperature regime during the vegetative growth period reduce the number of effective tillers in a commercial crop, farmers tend to use higher than recommended seed density to compensate for the loss of spikes. Although no concrete data have been generated, this practice may also be promoting WB infections in the non-epidemic years.

10.3.5 Varietal Diversification

Field observations indicate that later maturing varieties, in general, have less WB infection even in the epidemic years. Recently, Chávez and Kohli (2018a) reported that three later maturing varieties (Tbio Toruk, Tbio Iguazu and Tbio Sintonia) were considered moderately resistant even under artificially inoculated conditions. As a result, it is important to utilise a combination of early, intermediate and late varieties in different regions, in order that the flowering period does not coincide with disease-prone conditions and use maturity class as an escape mechanism from the disease.

In addition, there is plenty of variation in the levels of severity or susceptibility among different commercial varieties available to the farmer. As a result, the best approach to reduce WB infection is to use a combination of different varieties of diverse maturity groups representing distinct levels of resistance. Although such diversity does not guarantee lack of infection, it helps reduce the inoculum pressure and thus the damage caused to the field of a single variety.

10.3.6 Crop Rotation

Generally, crop rotation is considered an effective cultural practice to reduce the inoculum load of different disease-causing organisms. However, given the wide range of host species infected by *M. oryzae* in and around the field, an association between crop rotation and WB has not been observed in the Southern Cone region of South America. Crops, such as maize (*Zea mays*), soybean (*Glycine max*), mucuna (*Mucuna pruriens*) and crotalaria (*Crotalaria juncea*), have been used individually and in combination to rotate with wheat at the experimental station in Canindeyú, Paraguay. Although these rotations have led to better soil structure and fertility leading to higher grain yields, no differences have been observed in the frequency or severity of WB disease.

These experimental wheat plots are subjected to regular sprinkler irrigation to favour natural infections of WB. Under such conditions and depending on the level of resistance or susceptibility of a genotype, severe to moderate disease symptoms have been observed independent of the crop rotation followed. Similar results are seen in commercial fields, where a majority of the farmers rotate wheat crop with maize or soybean during the summer season and black oats during the winter.

It should be pointed out that lack of association between crop rotation and WB infection in the South American region should not be considered an argument not to adopt this useful practice elsewhere. The disease pressure and wide variability of hosts are probably a major reason for such incidences in South America, which may or may not be the case in other regions. On the contrary, an effective crop rotation system may be able to not only reduce the disease pressure on wheat, but also minimise the ability of *P. oryzae* to adapt to other hosts or local weed species.

10.4 UTILISATION OF RESISTANT CULTIVARS AND CHALLENGES

In an integrated approach to control WB epidemics, the development and utilisation of resistant or less susceptible varieties are considered the most effective and economical strategy.

10.4.1 Genetic Variability in Host and Fungus

In order to improve genetic resistance to WB, several programmes in Brazil and Paraguay have made it an important breeding goal. Early efforts to identify sources of genetic resistance to the disease were conducted by Barros et al. (1989), Igarashi (1990) and Goulart (1991) among others. While these evaluations led to the identification of a small number of promising lines with low WB infection under field conditions, most were found susceptible when artificially inoculated. Regional variability in cultivar performance, or their susceptibility under controlled conditions, allowed the identification of a restricted gene pool with WB resistance in bread wheat. Further, this genetic variation in resistance was often identified in old, tall and less productive cultivars which would not normally be considered in a breeding programme. The difficult task has been to develop adapted cultivars that combine a high and stable yield and quality traits with resistance to the relevant diseases and pests, including resistance to WB.

In order to ascertain the level and stability of resistance in older wheat genotypes, their evaluation with field samples of the pathogen were initiated. Although no pathogenic races of fungus causing WB have been reported in wheat, genetic variability has been observed and is considered important for the aggressiveness of *M. grisea* (Mehta & Baier, 1998; Urashima et al., 2004a). In order to account for this variability, Urashima et al. (2004a) studied 20 wheat cultivars from Brazil, using 72 monoconidial isolates of *M. grisea* from different regions of Brazil. Based on the compatible or incompatible reactions of leaf blast, they were able to identify 54 diverse virulence patterns, although 22% of the isolates showed similar virulence pattern. None of the wheat cultivars was found to be resistant to all isolates in this study. Measured by their disease index, the cultivars significantly differed in their degree of resistance.

10.4.2 Sources of Resistance

It is important to clarify that controlled studies utilising artificial leaf inoculations may not be truly representative of spike infections or resistance. Other studies have shown that leaf infection conducted at seedling stage, frequently does not translate to spike reaction or resistance (Arruda et al., 2005; Chavez et al., 2017). In a larger field study to identify WB resistance in 100 Brazilian wheat cultivars and breeding lines,

Prestes et al. (2007) recorded a large variability in the reaction to head blast infection ranging from 10% to 86% on average. None of the wheats displayed a high level of resistance to *M. grisea*. A group of 18 cultivars and breeding lines (including BRS 49, BRS 120, BRS 220 and IAPAR 53) were identified with lower infection ratings than the BR 18 cultivar, considered moderately resistant in Brazil. Similar studies conducted in Paraguay also confirm the susceptibility of most commercial varieties and advanced breeding lines, although significant differences in the reaction to WB were observed (Chavez et al., 2017; Chávez & Kohli, 2018a).

The identification of sources of resistance to WB in the Southern Cone region (Bolivia, Brazil and Paraguay) is solely based on field observations over years. Variable levels of infection, including severe epidemics, over the last 30 years have helped identify several wheat cultivars and advanced lines with moderate to high degrees of resistance to WB. In Bolivia, the cultivars Ichilo CIAT, Paragua CIAT and recently added Sausal CIAT, Motacu CIAT, Patuju CIAT and Urubu CIAT all have a moderately resistant to resistant reaction in the Santa Cruz region (Guzman, 2014). On the other hand, most of the cultivars released in Paraguay, except for Canindé 1 and Itapúa 75, are moderately susceptible to susceptible under field conditions (Kohli et al., 2010). It should also be pointed out that all durum wheats evaluated so far have been completely susceptible.

Four cultivars of the region (Sausal CIAT, Urubo CIAT [BO], CD 116 [BR] and Canindé 1 [PY]), identified as carrying moderate levels of field resistance, are all derivatives of the CIMMYT advanced line Milan (VS73.600/Mirlo/3/Bobwhite//Yoreme75/Trifon). It is believed that the VS 73.600 line, a progenitor of Milan, carries a 2NS chromosome segment from *T. ventricosa*, a source of WB resistance (Cruz et al., 2016). However, the genetics of the 2NS segment in wheat has still not been worked out. Kohli et al. (2011) warned that an increase in Milan-based resistant wheat cultivars in Bolivia, Brazil and Paraguay exposes them to a selection of new virulent pathotypes of the fungus and suggested that it be combined with other sources of resistance to keep it effective over the longer term. The other two moderately resistant cultivars, Motacu CIAT and Patuju CIAT, are synthetic-derived wheat advanced lines from CIMMYT.

A preliminary assessment of US wheat cultivars reported Postrock, JackPot, Overley, Jagalene, Jagger and Santa Fe cultivars show less than 3% infection. It is known that Jagger and Jagalene derive their resistance from the 2NS segment in wheat. It will be interesting to confirm the resistance of other sources mentioned above to help utilise them in the breeding programmes (Cruz et al., 2012).

Considering the performance of synthetic-derived wheats against WB, further exploration of *Aegilops tauschii* and other accessions should be a priority. Urashima and Kato (1994) screened a large number of *Aegilops* species against the wheat isolates of *Magnaporthe*. They identified only *Ae. tauschii* accessions to be resistant to all seven wheat isolates tested in the study, while *Ae. umbellulata* accessions were resistant to six of the isolates. Recently, Cazal et al. (2018) classified a group of 64 hexaploid synthetic wheats (*Triticum dicoccoides* × *T. tauschii*) against two isolates of *M. oryzae* and identified two of them with stable resistance to the disease under controlled conditions. All other species of *Aegilops*, carrying M or S genomes, were susceptible. In addition, there were differences in the level of resistance to WB among *Ae. tauschii* accessions collected from different geographical origins which can be further exploited to develop newer cultivars.

Additional wheat-based genes for resistance to WB have been named as *Rmg 2*, *Rmg 3* (from cultivar Thatcher) and *Rmg 6* (from *Triticum dicoccum*, KU120) by Zhan et al. (2008) and Tagle et al. (2015), respectively. While the germplasm carrying these genes have shown a moderate level of resistance to WB in Japan, they have not been tested in other regions to confirm their usefulness in the variety development programmes. In any case, they need to be exploited further to expand the actual narrow base of resistance to this disease.

10.4.3 Expediting Varietal Development

While conventional breeding programmes have been successful in improving the levels of genetic resistance in various diseases through repeated testing of breeding lines under induced and natural epidemic conditions, they definitely require long-term commitment and substantial investment to incorporate resistance in spike diseases such as WB. Further, WB resistance is only one of many traits required for a cultivar to be successful.

The assistance of molecular markers, a relatively recent tool, applied to enhance the efficiency of conventional breeding, especially for diseases such as WB and FHB, needs to be undertaken seriously. Except for the molecular marker to identify the presence of the 2NS segment in wheat, no other selection markers for WB resistance breeding have been published.

Currently, resistance breeding against WB is hampered from very limited genetic resources and lack of genomic knowledge. Given the wide variability reported in the fungus and the lack of adequate sources of resistance, the genetics of resistance has still not been worked out. Our studies have shown that cultivars showing complete resistance against a few isolates under controlled conditions in the glasshouse may or may not show field resistance in commercial cultivation (Chavez et al., 2017; Chávez & Kohli, 2018a). However, an effort is being made to combine different known sources of field resistance to develop newer sources with better and higher degrees of WB resistance.

10.5 CHEMICAL CONTROL MEASURES

10.5.1 Seed Treatment

The objective of chemical seed treatment is to completely eliminate the source of inoculum present in the seed and inhibit its transmission to the vegetative growth. Considering that this level of efficiency requires complete eradication of seed-associated fungi, it depends on the presence of pathogens, the fungi toxicity, the dose and the quality of the seed coverage by the fungicide in use.

In the Southern Cone region, the seed infection of WB is present simultaneously with the spot blotch fungus (*Bipolaris sorokiniana*). Considering that seed transmission of *M. oryzae* can result in 20%–30% of the WB incidence in the field, it is imperative to identify chemical seed treatments to reduce such infections.

In earlier studies, Goulart and Paiva (1996) were able to lower seed infection rates from 24% to 2% by using various fungicides and their combinations; however, they were not able to achieve totally clean seed. Since there are abundant sources of inoculum under field conditions, use of clean seed can definitely delay the onset of infection but cannot avoid an epidemic if the environmental conditions are favourable. In a field experiment utilising different concentrations of inoculum load on spikes, Pereira Gomes et al. (2017) observed loss of productivity and physiological quality of seeds in four Brazilian genotypes (BRS 264, VI 98053, CD 116 and CD 104). As a result, they recommended not using seed from any field with higher than 20% of WB infection.

10.5.2 Aerial Application of Fungicides and Management of Resistance to Fungicides

Aerial applications of fungicides to control spike diseases, viz. WB and *Fusarium* head blight, have not been very successful in general. This has probably less to do with the fungicidal efficiency and more to do with the poor application techniques and their timing. Although multiple applications between the heading and grain-filling stages are not sufficient to provide satisfactory control of these diseases, they certainly help reduce the damage (RCBPTT, 2017). Depending on the climatic conditions of the crop cycle and the varietal susceptibility or level of infection, the cost of multiple applications to control WB is generally not economical.

In Brazil, Cunha and Caierão (2015) recommend the cultural practices listed above (Section 10.3) as helpful means to reduce the WB damage in wheat and triticale. It further states that under environmental conditions that favour high disease pressure, the efficiency of fungicides to control WB is low and does not attain the objective but can help to reduce the damage. Cruz et al. (2015) mention that the control of foliar WB infections through fungicides is a valid alternative in order to reduce the number of *M. oryzae* conidia by 62%–77%, in comparison with two untreated check treatments. In a multi-location fungicide trial, Santana et al. (2013) reported lower infection and an increase in yield in those locations where the disease severity was moderate. However, under high disease pressure, the effect of fungicides was variable and they were not an economically viable disease control. They found that demethylation inhibitors (DMI fungicides, e.g. tebuconazole) were quite useful in reducing the incidence and severity of WB, but none of the tested fungicides provided an efficient level of control of WB. Similarly, Cruz (2013) also identified tebuconozole and metconazole to be effective in controlling WB, but the efficiency of their control was lower under warm and rainy weather conditions at the heading stage.

The extensive use of strobilurin fungicides (quinone-outside inhibitors [QoIs]) in Brazil has led to widespread resistance to QoI in *M. oryzae* populations associated with the evolution of the cytochrome b (*cyt b*) gene (Castroagudín et al., 2015). They identified the G143A mutation responsible for the QoI resistance and reported that its frequency in the wheat-infecting population of *M. oryzae* increased from 36% in 2005 to 90% in 2012, thereby requiring an urgent change of strategy to chemically control the disease.

TABLE 10.2
Effect of Chemical Control on the Grain Yield of Six Cultivars with Varying Levels of Resistance to Wheat Blast Disease, Santa Cruz, Bolivia, 2008–2009

Cultivar	WB reaction[a]	Grain yield (kg/ha) No control	Grain yield (kg/ha) w/Fungicide	Yield increase (%)
Azubi Ciat	S	1590	2899	182
BR 18	MS	2183	2700	124
Parapeti Ciat	MS	2634	3243	123
Paragua Ciat	MR	2595	3013	116
Achira Ciat	MR	2606	3040	117
Sausal Ciat	MR	3119	3453	111

Source: Unpublished data, Wheat Programme, Agricultural Research for the Tropics (CIAT).

[a] MR = moderately resistant; MS = moderately susceptible; S = susceptible.

It is important to mention that, in general, fungicides with different modes of action or their combinations are more useful to control spike diseases. The combination of fungicides from different chemical groups is necessary not only to broaden the scope of disease control but also to reduce the occurrence of cross-resistance as reported by Oliveira et al. (2015). They observed that all wheat isolates of *M. oryzae* sensitive to azoxystrobin were also sensitive to pyraclostrobin and vice versa. Those isolates carrying the G143 mutation resistant to azoxystrobin were also resistant to pyraclostrobin.

Although different fungicide combinations viz. epoxiconazole + pyraclostrobin and tebuconazole + trifloxystrobin were found to reduce the severity of WB and increase grain weight considerably, control of WB was more effective on leaves than spikes (Rocha et al., 2014). They also reported significant differences among varieties in the level of infection on the flag leaf but not the spike.

Chemical control of WB, though only partly effective, appears to be more useful when used in combination with a certain level of genetic resistance in a cultivar (Table 10.2).

10.5.3 Use of Resistance-inducing Chemicals

The inefficacy of the chemical control measures combined with the evolution of resistance in *M. oryzae* isolates has created a difficult situation for farmers to manage this disease, especially in an epidemic year. As a result, alternative control measures including the use of resistance-inducing chemicals have been proposed to control WB disease. It is believed that besides exerting less selection pressure on the pathogen population, these chemicals are environmentally friendly due to the rapid degradation of inducer molecules in an ambient environment. In a study conducted to evaluate the potential of resistance inducers to control WB disease,

Da Cruz et al. (2011) utilised jasmonic acid (AJ), deacetylated chitosan (QUI), potassium silicate (SP), acibenzolar-S-methyl (ASM), ethefon (ET) and potassium phosphite (PF) and compared them with tebuconazole fungicide (TE) and distilled water (AD) as control treatments. Three Brazilian cultivars (BR 23, BH 1146 and BRS 208) were sprayed with these chemicals 24 hours prior to inoculation with *M. oryzae*. While no significant difference between cultivars and treatments for the incubation period were found, an interaction between the control treatment and cultivar was observed for disease severity. However, any reduction in disease severity was conditioned by the cultivar, probably by its genetic resistance. The addition of TE and SP to the culture medium resulted in decreasing the mycelial growth of *M. oryzae* under laboratory conditions. The authors of this study emphasise the promising effect of PF and ET on host resistance and believe that the genetic resistance of a cultivar to WB can be enhanced by these chemicals.

10.6 BIOLOGICAL MANAGEMENT OF THE DISEASE

There is very little information on biological control as an effective option to reduce the losses caused by WB. Recently, Cabota et al. (2018) mentioned that two biocontrol agents (BCA), *Pseudomonas stutzeri* and *Trichoderma harzianum*, were not effective on plants infected by the most virulent isolate of *Magnaporthe*, most likely due to the differential response of *M. oryzae* isolates. However, when infected by a less virulent isolate, these BCAs triggered some biotic defence response in the host plants.

It is important to identify the pathogenic variability in populations of *M. oryzae* in order to detect the optimal antagonistic combinations that may be effective in their control. In a recent study, Dutta et al. (2018) combined three bacterial isolates of the *Bacillus* genus; two isolates of the *Bacillus amyloliquefaciens* and one of *B. subtilis*. The genomic analysis of these bacterial isolates revealed that they possess genes responsible for the biosynthesis of various antibiotics and enzymes with cell wall lytic functions. Some of these genes are involved in the biosynthesis of acetoin (acetolactate decarboxylase, alsSD), which induces systemic resistance in wheat plants. These authors believe that these bacterial strains and others can contribute to the control of WB disease in the future.

10.7 WEATHER-BASED EARLY WARNING SYSTEM FOR WB CONTROL

To a large extent, wheat yield in the Southern Cone region depends on environmental conditions and the technology under which it is grown. The multiplicity of foliar and spike diseases on wheat has forced farmers to use fungicides, in spite of the relatively lower yield potential. The chemical control of diseases has become an integral part of the wheat production system over the last three decades or so. Considering that a wheat crop provides only marginal returns in most years, any recommendation or model to reduce fungicide applications or their use in specific cases or areas/years will benefit the farming community.

While broad-spectrum fungicides are the farmers' first choice to apply against leaf and spike diseases, a decision support system can be very helpful to determine a specific risk assessment for WB infection or other diseases before or around wheat flowering. Thus, the decision to apply fungicide or not can be taken with more confidence. In order to support farmers against WB, Cardoso et al. (2008) used controlled temperatures and crop wetting periods in a growth chamber to develop a beta model to serve as a warning system for a WB epidemic. They used daily probability of infection to develop theoretical values for disease intensity which serve as a warning system. However, these authors also reported that minor fluctuations in the above factors, or others related to the host or pathogen, may distort the predictions under field conditions.

Recently, Fernandes et al. (2017) used a historical set (2001–2012) of disease and weather data from the northern Parana state, where the WB infection was first discovered, to identify weather patterns related to major outbreaks (2004 and 2009). They found that highly humid and warm weather during a 60-day period preceding wheat flowering were the key driving factors for inoculum buildup and airborne spores in a field as well as from other inoculum sources in the surroundings. Although they did not include agronomic management factors such as crop rotation, tillage type and cultivar resistance along with weather variables, they concentrated on estimating pathological variables viz. inoculum potential, spore cloud from models adapted to predict inoculum buildup and availability.

Based on daily climatic variables, temperature and relative humidity, and epidemic analysis, they identified a parameter called day favouring infection (DFI). Successful daily infection during a DFI was conditioned by inoculum potential (IP >30) and spore cloud (SPOR >0.4). Using these limits during the DFI, these authors were not only able to describe epidemic and non-epidemics years in the historical dataset, but also correctly predict the 2015 epidemic.

In order to help farmers in real-time decision-making, they have developed an interactive risk-mapping tool that collects actual weather data for the target area to warn of potential WB outbreaks. In addition to such a decision support system to help farmers online, their own knowledge regarding farming practices beneficial for WB control, viz. choice of resistant cultivars or change in the cropping system, are critical to limit fungicide applications in specific cases or years. Such an integrated approach to WB control can be applied in other regions where the disease is endemic or assess its epidemic potential in new regions.

10.8 TENTATIVE INTEGRATED PEST MANAGEMENT (IPM) MODULE FOR WHEAT BLAST

The restricted availability of genetic resistance to wheat blast disease in commercial cultivars and the characteristics of the pathogen, the wide host range and population variability require an integrated strategy of control to attenuate the effects of *M. oryzae*.

Good farming practices must start with an economically viable and sustainable crop rotation system to control the disease. It is important not only to curtail the adaptation or evolution of the pathogen on other grassy host species in and around the field, but also to reduce the initial inoculum load especially in the newer areas of the disease. It is essential to point out that the genetic resistance of a cultivar is of paramount importance in reducing the impacts of the disease. Our results show that use of cultivar having a moderate level of resistance immensely enhance the efficacy of cultural management and chemical control measures than using a susceptible cultivar.

The use of clean certified seed guarantees a healthy vegetative growth, although regular crop monitoring is essential to detect early leaf infections. Besides crop monitoring, a weather watch model developed to predict possible epidemics can be of enormous value to follow the evolution of the disease in the crop canopy. While leaf infections have not been found to cause much loss, except for the foliar area, they certainly become a key factor in the provision of inoculum at the heading or flowering stage of the crop. This is where preventive chemical control measures, using a combination of different families of fungicides, are critical to ensure minimum losses and high grain productivity.

The use of diversified cultivars with different crop cycles and levels of resistance to WB is crucial to diminish the agronomic losses that this disease can cause. Research data from the South American experience suggest that diversification of cultivars in terms of crop cycles and seeding dates prevents the flowering period from coinciding with infection-prone environmental conditions (rains or high humidity), thereby avoiding the disease. It is also important to broaden the genetic base of cultivars for resistance, including different reactions to WB, in order to avoid the risks of overcoming WB through pathogen evolution or variability. Although some biological control methods have been explored and proposed in the literature as promising alternatives, no concrete recommendations of their applicability on a large scale have been reported or tested to date. Once their efficacy can be proven under field conditions, these can certainly form an integral part of the WB disease management strategy.

A graphical representation of an integrated control scheme to reduce the losses caused by WB in the South American region is presented in Figure 10.1. Although there may be minor variations in its use among countries or regions, with respect to the component importance assigned in Figure 10.1, they have all been essential in the fight against WB for over three decades.

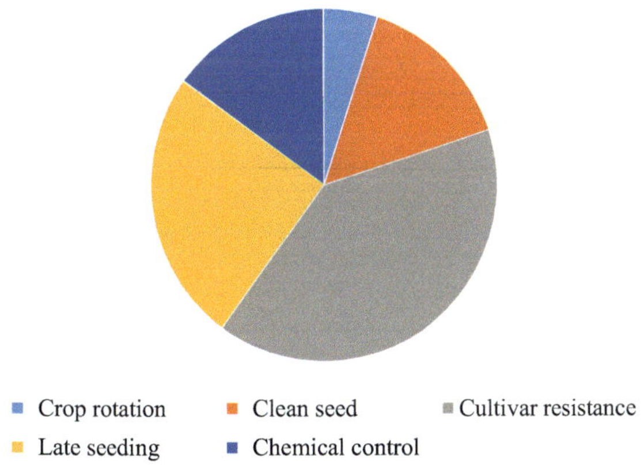

FIGURE 10.1 Components of an integrated strategy to control WB in the South American region and their importance.

10.9 FUTURE NEEDS FOR RESEARCH

In spite of the fact that WB was identified over 30 years ago, very few studies describe the pathogen variability in different geographical areas and its relationship with the genetic base of resistance in wheat. Buerstmayer et al. (2017) believe that current resistance breeding against WB is hampered by very limited genetic resources and lack of genomic knowledge. Although there are many studies on *Magnaporthe* species attacking wheat and other hosts in and around the field in the literature, the mechanism of spore dissemination or distribution to create a high load of inoculum in an epidemic year is not well understood.

Gene identification and their functional analysis are essential components to widen the scope of resistance and its transfer into high-yielding cultivars; however, there is significant lack in the identification of a diversified genetic base of resistance in wheat and allied species. It is also imperative to enhance the identification of molecular markers linked to the resistance loci and/or pathogen diversity in order to screen the segregating populations under development. To date, no reliable selection markers for WB resistance breeding have been published.

In addition to the genetic control of WB, designing optimal integrated strategies, including agronomic (crop rotation) practices, and chemical control of the disease need to be clearly defined for different geographical regions. The development of novel fungicides, safe for the environment and for humans, which have better disease control efficacy would also be highly desirable. Given the unpredictable spread of WB and its potential to cause damage, refining real-time, site-specific decision support systems is urgent and warrants further investment.

REFERENCES

Anjos, J. R. D., D. B. Da Silva, M. J. Charchar and G. Rodrigues. "Ocorrência de Brusone (*Pyricularia grisea*) Em Trigo e Centeio Na Região Dos Cerrados Do Brasil Central." *Pesquisa Agropecuária Brasileira* 31, no. 1 (1996): 79–82.

Arruda, M. A, C. R. Bueno, K. C. Zamprogno, N. A. Lavorenti and A. S. Urashima. "Reação Do Trigo à *Magnaporthe grisea* Nos Diferentes Estádios de Desenvolvimento." *Fitopatologia Brasileira* 30, no. 2 (2005): 121–26.

Barros, B. C., J. C. Felicio, C. E. O. Camargo and A. W. P. Ferreira Filho. "Reação de Cultivares de Trigo a Brusone em Condições de Campo." *Summa Phytopathologica* 15 (1989): 21.

Buerstmayer, H., V. Mohler and M. Kohli. "Advances in Control of Wheat Diseases: Fusarium Head Blight, Wheat Blast and Powdery Mildew." In *Achieving Sustainable Cultivation of wheat Volume 1.* (Edited by Peter Langridge) Burleigh Dodds Science Publishing Limited, UK. Chapter 13, 2017. doi:10.19103/AS.2016.0004.21.

Cabota, C., R. Boscha, S. Martosc, C. Poschenriederc and A. Perelló. "Salinity is a Prevailing Factor for Amelioration of Wheat Blast by Biocontrol Agents." *Biological Control* 125 (2018): 81–9. doi:10.1016/j.biocontrol.2018.07.003.

Callaway, E. "Devastating Wheat Fungus Appears in Asia for First Time." *Nature* 532, no. 7600 (2016): 421–2. doi:10.1038/532421a.

CAO, CÁMARA AGROPECUARIA DEL ORIENTE. *Evaluación De Desempeño Del Sector Agropecuario En El Departamento de Santa Cruz.* Santa Cruz: CÁMARA AGROPECUARIA DEL ORIENTE, 2014.

Cardoso, C. A. A., E. Melo Reis and É. N. Moreira. "Development of a Warning System for Wheat Blast Caused by *Pyricularia grisea*." *Summa Phytopathologica* 34, no. 3 (2008): 216–21. doi:10.1590/S0100-54052008000300002.

Castroagudín, V. L, P. C. Ceresini, S. C de Oliveira et al. "Resistance to QoI Fungicides is Widespread in Brazilian Populations of the Wheat Blast Pathogen *Magnaporthe oryzae*." *Phytopathology* 105, no. 3 (2015): 284–94. doi:10.1094/PHYTO-06-14-0184-R.

Castroagudín, V. L., S. I. Moreira, D. A. S. Pereira et al. "*Pyricularia Graminis-Tritici*, a New Pyricularia Species Causing Wheat Blast." *Persoonia – Molecular Phylogeny and Evolution of Fungi* 37 no. 1 (2016): 199–216.doi:10.3767/003158516X692149.

Cazal, C. C., A. R. Chávez, Y. M. Reyes, M. M. Kohli and P. E. Pérez. "Evaluation of Synthetic Hexaploid Wheats for Resistance to Wheat Blast Disease." *Mexican Journal of Phytopathology* 37, no. 1 (2018): 1–15. doi:10.18781/R.MEX.FIT.1807-5.

Chávez, A. and M. M. Kohli. "Alternative Hosts of *Magnaporthe grisea* of Wheat in Paraguay." *Investigación Agraria* 17, no. 1 (2015): 54–9.

Chavez, A., C. C. Cazal and M. M. Kohli. "Difference in the Reaction to *Pyricularia oryzae* of Wheat Materials in the Vegetative and Reproductive Stages." *Investigación Agraria* 19, no. 1 (2017): 56–63. doi:10.18004/investig.agrar.2017.junio.56-63.

Chávez, A. R. and M. Kohli. "Patogenicidad de *Magnaporthe oryzae* En Variedades y Líneas de Trigo Cultivadas En Paraguay." *Revista Mexicana de Fitopatología, Mexican Journal of Phytopathology* 36, no. 2 (2018a): 276–86.doi:10.18781/R.MEX.FIT.1712-3.

Chávez, A. R. and M. M. Kohli. "Identificación de Hospedantes Alternativos de Magnaporthe Sp. En Campos de Trigo de Paraguay." In *Pyricularia de Trigo En Paraguay. Compendio de Investigación*, edited by Chávez, A. R. and M. M. Kohli 18–25. Asunción: Cámara Paraguaya de Exportadores y comercializadores de cereales y oleaginosas, 2018b.

CONAB, Companhia Nacional de Abastecimento. 2015. https://www.conab.gov.br/conteudos.php?a=1252&t=&Pagina_%2520objcmsconteudos=%25203#A _objcmsconteudos.

Couch, B. C. and L. M. Kohn. "A Multilocus Gene Genealogy Concordant with Host Preference Indicates Segregation of a New Species, *Magnaporthe oryzae*, from *M. grisea*." *Mycologia* 94, no. 4 (2002): 683–93. doi:10.2307/3761719.

Cruz, C. D. *Wheat Blast: Quantitative Pathway Analyses for the Triticum Pathotype of* Magnaporthe oryzae *and Phenotypic Reaction of U.S. Wheat Cultivars.* Kansas State University, Manhattan, KS, 2013.

Cruz, C. D and B. Valent. "Wheat Blast Disease: Danger on the Move." *Tropical Plant Pathology* 42, no. 3 (2017): 210–22. doi:10.1007/s40858-017-0159-z.

Cruz, C. D., J. Kiyuna, W. W. Bockus, T. C. Todd, J. P. Stack and B. Valent. "*Magnaporthe oryzae* Conidia on Basal Wheat Leaves as a Potential Source of Wheat Blast Inoculum." *Plant Pathology* 64, no. 6 (2015): 1491–8. doi:10.1111/ppa.12414.

Cruz, C. D., W. W. Bockus, J. P. Stack et al. "Preliminary Assessment of Resistance Among U.S. Wheat Cultivars to the Triticum Pathotype of *Magnaporthe oryzae*." *Plant Disease* 96 (2012): 1501–5. doi:10.1094/PDIS-11-11-0944-RE.

Cruz, C. D, W. W. Peterson, W. W. Bockus et al. "The 2NS Translocation from *Aegilops Ventricosa* Confers Resistance to the Triticum Pathotype of *Magnaporthe oryzae*." *Crop Science* 56 (2016): 990–1000. doi:10.2135/cropsci2015.07.0410.

Cruz, M. da, A. P. C. Diniz, F. A. Rodrigues and E. Gonçalves De Barrios. "Aplicação Foliar de Produtos Na Redução Da Severidade Da Brusone Do Trigo." *Tropical Plant Pathology* 36 (2011): 424–8.

Cunha, G. R. Da and E. Caierão. "Informações Técnicas Para Trigo e Triticale – Safra 2015." In *VIII Reunião Da Comissão Brasileira de Pesquisa de Trigo e Triticale*, Canela, Brazil 229. EMBRAPA, 2015.

Dutta, S., Z. Musrat, M. Surovy et al. "Genomic Analyses Reveal that Biocontrol of Wheat Blast by *Bacillus* spp. may be Linked with Production of Antimicrobial Compounds and Induced Systemic Resistance in Host Plants." 2018. doi:10.6084/m9.figshare.5852661.v1.

Embrapa Trigo. "Brusone Do Trigo é Identificada Em Lavouras Do RS. Agron Beta, 2014." 2014. https://www.agron.com.br/imprimir.html?topico=041251.

Fernandes, J. M. C., M. Nicolau, W. Pavan et al. "A Weather-Based Model for Predicting Early Season Inoculum Build-up and Spike Infection by the Wheat Blast Pathogen." *Tropical Plant Pathology* 42, no. 3 (2017): 230–37. doi:10.1007/s40858-017-0164-2.

Gladieux, P., B. Condon, S. Ravel et al. "Gene Flow Between Divergent Cereal- and Grass-Specific Lineages of the Rice Blast Fungus *Magnaporthe oryzae*." *American Society for Microbiology* 9, no. 1 (2018): e01219–17. doi:10.1128/mBio.01219-17.

Goulart, A. *Doencas Do Trigo Em Mato Grosso Do Sul*. Dourados: EMBRAPA. Unidade de Execucao de Pesquisa de Ambito Estadual de Dourados (MS), 1991.

Goulart, A, F. A. Paiva and A. N. Mesquita. "Ocorrência de Brusone (*Pyricularia oryzae*) Do Trigo (*Triticum aestivum*) Em Mato Grosso Do Sul." *Fitopatologia Brasileira* 15, no. 1 (1990): 112–4.

Goulart, A, F. A Paiva, A. N Mesquita and de A. G Moraes. "Transmissão de *Pyricularia oryzae* Cav. Pela Semente de Trigo (*Triticum aestivum* L.)." In *REUNIÃO DA COMISSÃO CENTROSUL BRASILEIRA DE PESQUISA DE TRIGO – Cornélio Procópio*. Resultados de Pesquisa de Com Trigo – 1989, 170, 1989.

Goulart, A. C. P. and F. A. Paiva. "Efeito Do Tratamento de Sementes de Trigo (*Triticum aestivum* L.) Com Fungicidas Para Controle de Heliminthosporium Sativum e *Pyricularia oryzae*." *Ciência e Agrotecnologia* 20, no. 1 (1996): 25–30.

Guzman, E. R. *Mejoramiento de Trigo En Santa Cruz*. Santa Cruz: Centro de Investigación Agrícola Tropical (CIAT), 2014.

Igarashi, S. "Update on Wheat Blast (*Pyricularia oryzae*) in Brazil." In *Wheat for the Nontraditional Warm Areas*, edited by D. A. Saunders, 480–3. Mexico, DF: CIMMYT, 1990.

Igarashi, S., C. M. Utiamada, A. H. Kasuma and R. S. López. "*Pyricularia* sp em Trigo. 1. Ocurrencia de Pyricularia sp No Estado Do Paraná." *Fitopatología Brasilera* 11 (1986): 351–2.

Inoue, Y., T. T. P. Vy, K. Yoshida et al. "Evolution of the Wheat Blast Fungus through Functional Losses in a Host Specificity Determinant." *Science* 357, no. 6346 (2017): 80–3. doi:10.1126/science.aam9654.

Islam, M. T., D. Croll, P. Gladieux et al. "Emergence of Wheat Blast in Bangladesh was Caused by a South American Lineage of *Magnaporthe oryzae*." *BMC Biology* 14, no. 1 (2016). doi:10.1186/s12915-016-0309-7.

Klaubauf, S., D. Tharreau, E. Fournier et al. "Resolving the Polyphyletic Nature of Pyricularia (*Pyriculariaceae*)." *Studies in Mycology* 79 (2014): 85–120. doi:10.1016/j.simyco.2014.09.004.

Kohli, M., Y. R. Mehta, E. Guzman, L. de Viedma and L. E. Cubilla. "Pyricularia Blast a Threat to Wheat Cultivation." *Czech Journal of Genetics and Plant Breeding* 47, SPEC. ISSUE 1 (2011): 130–4. doi:10.17221/3267-CJGPB.

Kohli, M. M., L. de Viedma and L. E. Cubilla. *Guía Para La Producción de Trigo. Fortalecimiento de La Investigación y Difusión Del Cultivo de Trigo En Paraguay*. Paraguay: M. M. Kohli, Lidia de Viedma, L. E. Cubilla, 2010. https://issuu.com/py finances/docs/guia_para_la_produccion_de_trigo

Malaker, P. K., N. C. D. Barma, T. P. Tiwari et al. "First Report of Wheat Blast Caused by *Magnaporthe oryzae* Pathotype Triticum in Bangladesh." *Plant Disease* (2016), Scientific Societies, PDIS-05-16-0666-PDN. doi:10.1094/PDIS-05-16-0666-PDN.

Marangoni, M. S., M. P. Nunes, J. N. Fonseca and Y. R. Mehta. "Pyricularia Blast on White Oats: A New Threat to Wheat Cultivation." *Tropical Plant Pathology* 38, no. 3 (2013): 198–202. doi: 10.1590/S1982-56762013005000004.

Mehta, Y. R. and A. Baier. "Variação Patogênica Entre Isolados de *Magnaporthe grisea* Atacando Triticale e Trigo No Estado Do Paraná." *Summa Phytopathologica* 24, no. 2 (1998): 119–25.

Mehta, Y. R., C. R. Riede, L. A. C. Campos and M. M. Kohli. "Integrated Management of Major Wheat Diseases in Brazil: An Example for the Southern Cone Region of Latin America." *Crop Protection* 11, no. 6 (1992): 517–24. doi:10.1016/0261-2194(92)90168-5.

Oliveira Coelho, M. A. de, G. A. Montan Torres, P. R. Cecon and F. Martins Santana. "Sowing Date Reduces the Incidence of Wheat Blast Disease." *Pesquisa Agropecuaria Brasileira* 51, no. 5 (2016): 631–7. doi:10.1590/S0100-204X2016000500025.

Oliveira, S. C. De, V. L. Castroagudín, J. L. Nunes Maciel, D. A. S. Pereira and P. C. Ceresini. "Resistência Cruzada Aos Fungicidas IQo Azoxistrobina e Piraclostrobina No Patógeno Da Brusone Do Trigo *Pyricularia oryzae* No Brasil." *Summa Phytopathologica* 41, no. 4 (2015): 298–304. doi:10.1590/0100-5405/2072.

Pereira Gomes, D., V. S. Rocha, O. L. Pereira and M. Alves De Souza. "Damage of Wheat Blast on the Productivity and Quality of Seeds as a Function of the Initial Inoculum in the Field." *Journal of Seed Science* 39, no. 1 (2017): 66–74. doi:10.1590/2317-1545v39n1172688.

Picinini, E. C. and J. M. C. Fernandes. "Ocorrência Da Brusone (*Pyricularia oryzae*) Em Lavouras Comerciais de Trigo (*Triticum aestivum*) No Estado Do Rio Grande Do Sul." *Fitopatologia Brasileira* 14, no. 2 (1989): 126.

Pieck, M. L., A. Ruck, M. Farman et al. "Genomics-Based Marker Discovery and Diagnostic Assay Development for Wheat Blast." *Plant Disease* 101, no. 1 (2017): 103–9. doi:10.1094/PDIS-04-16-0500-RE.

Prabhu, A. S., M. C. Filippi and N. Castro. "Pathogenic Variation among Isolates of *Pyricularia oryzae* Affecting Rice, Wheat, and Grasses in Brazil." *Tropical Pest Management* 38, no. 4 (1992): 367–71. doi:10.1080/09670879209371729.

Prestes, A. M., P. F. Arendt, P. L. Fernandes and J. M. C. Scheeren. "Resistance to *Magnaporthe grisea* Among Brazilian Wheat Genotypes." In *Wheat Production in Stressed Environments. Developments in Plant Breeding*, 12th ed. edited by H. T. Buck, J. E. Nisi and N. Salomón, 119–24. Dordrecht: Springer, 2007. doi:10.1007/1-4020-5497-1_16.

PROTRIGO. *Programa Nacional de Investigación y Transferencia de Tecnología Para El Cultivo de Trigo Fase II*, Santa Cruz, Bolivia. 2002, 15–20.

RCBPTT, Reunião da Comissão Brasileira de Pesquisa de Trigo e Triticale. *Informações Técnicas Para Trigo e Triticale – Safra 2017.* Londrina, PR: Embrapa, 2017.

Rocha, J., R. Santos de Carvalho, A. J. Badaró Pimentel, G. Ribeiro and M. Alves de Souza. "Eficiência de Fungicidas No Controle Da Brusone Em Trigo." *Summa Phytopathologica* 40, no. 4 (2014): 347–52. doi:10.1590/0100-5405/1937.

Santana, F. M., J. L. N. Maciel, Douglas Lau et al. *Comunicado 328 Técnico.* Passo Fundo, RS: Empresa Brasileira de Pesquisa Agropecuária, 2013.

Tagle, A., I. Chuma and Y. Tosa. "*Rmg7*, a New Gene for Resistance to Triticum Isolates of *Pyricularia oryzae* Identified in Tetraploid Wheat." *Phytopathology* 105, no. 4 (2015): 495–9. doi:10.1126/science.49.1259.174.

Tanaka, M. A. S., M. L. B. Alves and J. O. M. Menten. "*Pyricularia* Spp. de Trigo e Arroz: II-Transmissão Através de Sementes Inoculaladas." *Fitopatologia Brasileira* 13 (1988): 106.

Toledo, J. and R. Escobar. *Piricularia o Bruzone Del Trigo.* Santa Cruz, CA: Centro de Investigación Agrícola Tropical (CIAT), 2002.

Urashima, A. S. and H. Kato. "Varietal Resistance and Chemical Control of Wheat Blast Fungus." *Summa Phytopathologica* 20, no. 2 (1994): 107–12.

Urashima, A. S., S. Igarashi and H. Kato. "Host Range, Mating Type, and Fertility of *Pyricularia grisea* from Wheat in Brazil." *Plant Disease* 77 (1993): 1211–6. http://ukpmc.ac.uk/abstract/AGR/IND20368077.

Urashima, A S., N. A. Lavorent, A. C. P. Goulart and Y. R. Mehta. "Resistance Spectra of Wheat Cultivars and Virulence Diversity of *Magnaporthe grisea* Isolates in Brazil." *Fitopatologia Brasileira* 29, no. 5 (2004a): 511–8. doi:10.1590/S0100-41582004000500007.

Urashima, A. S., T. Dias Martins, C. R. N. C. Bueno, D. B. Favaro, M. A. Arruda and Y. R. Mehta. "Triticale and Barley: New Hosts of *Magnaporthe grisea* in São Paulo, Brazil. Relationship with Blast of Rice and Wheat." In *Rice Blast: Interaction with Rice and Control*, edited by Shinji Kawasaki, 251–60. Dordrecht: Springer, 2004b.

Wang, H., M. Bo Qu and S. Wei. "First Report of a Leaf Blight Caused by *Pyricularia Pennisetigena* on *Cenchrus Pauciflorus* in China." *Plant Disease* 103, no. 2 (2019): 370.

Yasuhara-Bell, J., K. F. Pedley, M. Farman, B. Valent and J. P. Stack. "Specific Detection of the Wheat Blast Pathogen (*Magnaporthe oryzae* Triticum) by Loop-Mediated Isothermal Amplification." *Plant Disease* 102, no. 12 (2018): 2550–9.

Zhan, S. W., S. Mayama and Y. Tosa. "Identification of Two Genes for Resistance to Triticum Isolates of *Magnaporthe oryzae* in Wheat." *Genome* 51, no. 3 (2008): 216–21. doi:10.1139/G07-094.

Index